CW01267336

ABOUT THE AUTHOR

William Jones, MTECH, MENG, MIEI, works as a manufacturing consultant, providing training and consultancy services to a cross-section of industry. He qualified from the University of Limerick with a Mechanical Engineering degree and a Master of Technology degree in Advanced Manufacturing Technology. He has worked with a variety of world-leading companies including Digital Equipment Corporation, Apple Computers Ltd., General Instrument, Elsevier Scientific Publishers and Motorola. In industry, he has worked on a range of projects including JIT implementation, ISO 9000/ QS 9000, Equipment Maintenance Improvement, Computer Simulation, amongst others.

The Handbook of Modern Manufacturing Techniques

William Jones

Oak Tree Press
Dublin

Oak Tree Press
Merrion Building
Lower Merrion Street
Dublin 2, Ireland

© 1998 William Jones

A catalogue record of this book is
available from the British Library

ISBN 1-86076-070-8

All rights reserved. No part of this publication may be reproduced or transmitted in any form or by any means, including photocopying and recording, without written permission of the publisher and the individual contributors. Such written permission must also be obtained before any part of this publication is stored in a retrieval system of any nature. Requests for permission should be directed to
Oak Tree Press, Merrion Building,
Lower Merrion Street, Dublin 2, Ireland.

Printed in Ireland by Colour Books Ltd.

Contents

Acknowledgements vii

Preface xi

Introduction 1

Chapter 1
Operator Empowerment 9

Chapter 2
Total Quality Management Tools 39

Chapter 3
Quality Systems: ISO 9000 and QS 9000 89

Chapter 4
Equipment Engineering and Maintenance 129

Chapter 5
Just-in-Time Production 163

Chapter 6
Computer Simulation 215

Chapter 7
Process Reengineering 239

Chapter 8
Automation Technologies 257

Chapter 9
Project Management 293

Appendices 319
Bibliography 325
Index 331

ACKNOWLEDGEMENTS

I would like to acknowledge the support and help received from many people. Thanks to my family who initially encouraged me to undertake this project and reviewed early drafts. I would also like to thank the staff at Oak Tree Press, especially Brian Langan and David Givens, who invested their time and effort in developing this book. The ideas described in the book come from working on a variety of projects and studying at the University of Limerick. Thanks to all the people with whom I have worked and learned. Finally, I'd like to thank Patrick Mulcahy for a solid introduction to science.

*Dedicated to my wife Edel, without whose help
this book would not be finished*

PREFACE

In the coming decade, competition is set to increase as Europe becomes more cohesive and new markets are opened in Eastern Europe and Asia. Survival in industry will depend on exploiting opportunities for improving operational effectiveness through the use of modern manufacturing techniques. As managers wrestle with this increasingly competitive and fast-changing environment, they need access to a practical guide to these new techniques. This book presents these modern manufacturing techniques in an easy-to-read, concise format with pragmatic advice on implementation.

This book is written by a manufacturing engineer who has worked with several Fortune 500 companies and witnessed the benefits derived from modern manufacturing techniques. Each of the techniques are described from a practical perspective and advice is provided on avoiding common pitfalls. Only the minimum of manufacturing or mathematical experience is assumed. Each chapter is a self-contained description of the various techniques in use. The reader is introduced to the concepts through a series of steps: introduction; definition; suitability; and analysis of the technique.

The book aims to encourage implementation of the new techniques by providing a simple, concise description of the fundamentals of each. The design of the book makes it easy to read and a useful tool for manufacturing managers and engineers. I hope you find in its contents ideas and concepts that can be applied within your work environment.

Critiques, questions or suggestions regarding this book should be addressed to the author: William Jones, 34 The Innings, Lower Rathmines Road, Dublin 6.

INTRODUCTION

In the 1990s, manufacturing industries have undergone a period of transition as companies develop global perspectives in their businesses. Global competitiveness means that companies from Japan, America, Europe and elsewhere are all competing for the same market. There was a time when companies concentrated their manufacturing resources within their national boundaries, but today, manufacturing is a mobile business component and is located wherever strategic advantage can be gained.

Many of the mass manufacturing industries have moved to developing countries in Asia where low-cost labour minimises manufacturing costs. The trend in moving manufacturing facilities to these production sites began in the 1970s, but has become an epidemic in the 1990s, especially with the availability of an almost limitless supply of labour in China.

In order to compete with production facilities in developing countries, Western manufacturers must compensate for its high-cost base by excelling in all other areas of manufacturing. The processes by which people, raw materials and equipment are co-ordinated must be fine-tuned to eliminate all sources of excess and waste. In the 1980s, it was sufficient to produce goods of high quality in order to sustain market share. Japanese excellence in manufacturing quality goods led to explosive growth in their automotive, consumer electronics and steel industries. Recently, other countries along the Asian rim have replaced their image as low quality manufacturers by following Japan's lead. In today's environment, the competition war is fought over a range of metrics, including delivery lead-time to customers, aggressive pricing of products, diverse product range, among others. Best-in-class performance in each of these areas depends on the business the company is in. A consumer goods manufacturer may consider a lead-time of one week as world class, while an industrial parts manufacturer might strive for four weeks lead-time as a best-in-class goal. Typical world class performance levels would be:

- Delivery reliability   Greater than 98%
- Net inventory turns   30–50 per year
- Final inspection   100%
- Quality level   60 reject parts per million
- Cost of scrap   2% of sales
- Suggestions per employee   25
- Suggestions implemented   75%
- Net cost reduction   10%
- D.L. productivity   100% to plan
- Equipment downtime   less than 1%

The way to achieve success in each of these areas is to return to basics. The know-how is available and the techniques are proven. The challenge is to adapt the techniques to the individual requirements of each manufacturing facility.

It can sometimes be difficult to fully appreciate the seriousness of competition, especially when production personnel often struggle from meeting one end-of-month target to the next. The decline of a company is generally a gradual process, especially for multinational companies. From time to time, a customer is lost, or capital for investment in new equipment is not as readily available. In 1972, IBM and General Motors were ranked one and four respectively in the Fortune 500 listing for that year; in 1992, neither of these companies was in the top 20. The average lifespan of a typical multinational is 40 to 50 years, which means that of the companies featured in the current fortune 500, about one-third will not be present in their current form in the year 2010. They will have merged, been broken up or gone out of business. For smaller companies, the prospects are worse, since the life expectancy of the average European and Japanese company is less than 13 years.

The pace of development in manufacturing means that it can be either a competitive tool for defending market share or a sluggish entity dragging down the business. The globalisation

Introduction

of companies means that competition is increasing, but also, the integration of countries within the EU, the emergence of central Europe, the availability of new technologies means that there are many opportunities for growth. Those companies who are able to optimise their systems will be able to react to these new opportunities. They will ensure competitiveness by maximising the efficiency of the manufacturing process.

FIGURE 1: COMPONENTS OF MANUFACTURING

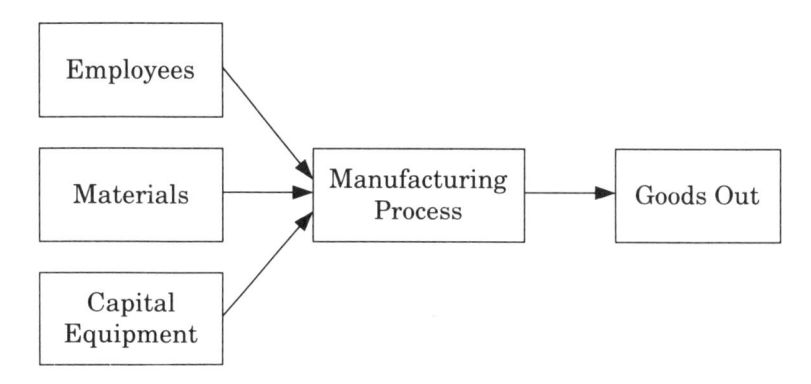

```
  Employees ┐
            │
  Materials ┼──► Manufacturing Process ──► Goods Out
            │
  Capital   ┘
  Equipment
```

Employees are probably the most important resource available to the company. They are the innovators who can overcome the many problems associated with manufacturing. Operators are the integrating factor that links materials and equipment within the process to produce the goods. They can stifle and even strangle progress within the company, or they can innovate, make suggestions, help in implementation and be the difference between success and failure. The traditionalist approach to manufacturing was based on the ideas of Frederick Taylor and scientific management. Taylor advocated separating the management of work from its execution. Industrial engineering techniques were developed which measured the work content in different jobs and provided management with the information needed to direct operators. Operators were considered as resources similar to machines. They performed a physical task and were paid accordingly. Taylor's ideas have led to much antagonism between management and operators in manufacturing. When automation technology became available in the 1950s and 1960s, it was generally perceived as a means for replacing obstreperous operators on the production floor, as well as improving productivity. In the 1990s, many companies are adopting different approaches to operator management. Operators are perceived as being able to contribute with their heads as well as their hands. New working structures have been developed which encourage operator participation in the management of their work. These are the people who have a keen awareness of manufacturing problems and potential solutions, because they are in direct contact with the problem every working day. They are an engineering and management resource able to provide an insight into the problem and help in implementing solutions. Instead of being an extension of machinery, they are now an extension of management, working to improve the performance in a multitude of different ways.

In Matsushita Electric Industries, the average number of improvement suggestions per employee per year is 45, with an implementation rate higher than 75 per cent. This results in a lot of improvement ideas being implemented every year. This is one of the ways to maximise the contribution that employees can make. Companies that provide employees with an opportunity to contribute are generally surprised and impressed with the response. The whole workforce becomes mobilised and tuned in to the needs of the organisation. Ownership is pushed down the hierarchical chain to where people have the most experience and sometimes even the best ideas and solutions to problems.

The philosophy of mobilising the workforce is reinforced by the development of total quality management techniques. These improvement techniques are designed for use in team environments where participants can exchange ideas, evaluate solutions and control problems. TQM tools are easy to use and provide a step-by-step guide to systematically bringing the production process under control. TQM is a continuous improvement methodology where the team strives to eliminate all sources of variation for the production process. The methods and philosophy provide a framework, enabling people to apply themselves to improving manufacturing performance.

Manufacturing efficiency depends on the effective utilisation of all resources, including equipment. As automation technologies become more prevalent, the success of the facility is dependent on management's ability to introduce and control equipment performance. A survey by Ernst and Young (1989) indicated that top management's poor understanding of advanced manufacturing technologies was the main obstacle to implementation of these technologies. In many cases, investment in technology is left until problems become crises, resulting in investment decisions being made in haste. Investment in technology requires a long-term perspective to ensure that new technologies complement each other and that the technical engineering expertise is available in-house to sustain them. Making the right choice of equipment is only the first step; making it work effectively is the second step in equipment management. Unreliable equipment results in poor product quality, poor delivery reliability, and excessive expenditure in spares and technical support resources. Again, the Japanese are leading the way in establishing effective equipment maintenance systems. They have developed the concept of Total Productive Maintenance (TPM) that combines traditional maintenance concepts with new ones. TPM trains production operators in how to look after machines, how to keep them clean and how to make minor repairs. In so doing, they are the first line of defence in maintaining properly functioning machines. TPM encourages technicians to work closely with operators in analysing the causes of breakdowns and implementing permanent fixes for repetitive problems. This approach to equipment management prolongs the lifetime of equipment and increases the company's return on investment. It makes it easier to continue a policy of investment in modern technology.

The other key resource in manufacturing is materials. The aim of manufacturing is to add value to raw materials and ship the goods to the customer in the shortest possible timespan. Working capital is minimised by minimising the inventory levels on hand at any time. Just-in-Time manufacturing revolutionised management's perception of inventory in the 1980s. The traditional perception of inventory was that high levels led to high utilisation rates for equipment, since the probability of material shortages were much reduced. This concept suited a time when labour and equipment were the main components of product cost. However, in many industries, materials are the main cost factor affecting unit cost and, as such, high material utilisation has replaced equipment and people utilisation as the key factor for controlling costs. Just-in-Time techniques are geared towards moving raw materials through the manufacturing processes in the shortest timespan. They achieve this goal by linking each of the processes so that production is co-ordinated between workstations. Each workstation produces only what is requested by subsequent workstations in the process. Using this simple philosophy, each of the workstations is controlled by their "customer" workstation downstream in the process. The supplier–customer links within the factory are extended to the factory's suppliers and distribution centre to minimise inventory levels all along the production chain. Just-in-Time can be difficult to implement, since it requires close co-ordination between workstations within the factory and between the factory and its suppliers/distribution centre. However, the benefits derived are a clear justification for the effort required in its implementation.

The pace of change in today's world means that the manufacturing manager has to react fast to the industrial environment. The manager has to stay abreast of new management trends and new technological developments. Effective decision-making can enable the factory to gain competitive advantage, while poor decision-making means resources are wasted and options reduced. There are a number of support tools and techniques available for decision-making, three of which are computer simulation, process reengineering and project management techniques. Computer simulation software has become relatively inexpensive in recent years and need no longer be considered a heavy investment for the company. Computer simulation enables the manager to predict the effect of decison-making. The manager constructs a computerised representation of the manufacturing facility, describes how decisions affect the process (e.g. new machine or new product variant) and describes random characteristics associated with the decision (e.g. expected breakdowns for the new machine or expected demands for the new product variants), and the software estimates the effect of random factors on the overall process performance. The sensitivity of the process to variations — in demand or breakdown or whatever variable is appropriate to the test — can be easily determined and management can calculate their risk exposure.

Process reengineering means re-evaluating the methods and systems used within the manufacturing facility. The processes for performing the various tasks associated with manufacturing, from order acceptance to delivery of goods to the customer, are scrutinised in light of the capability of information technology. Many of the processes operating in companies were designed before the advent of computer technologies and are therefore out of date. Companies can gain competitive advantage by recognising the potential of information technologies and redesigning processes and systems to incorporate these.

Project management techniques have been well known for some time, but it is only recently that project management software, which automatically makes all the calculations associated with PERT charts, Gantt chart and resource allocation, has become available at low cost. The availability of software to perform the maths has rekindled interest in project management techniques. Systematic project management helps to avoid oversight and optimise resource utilisation during implementation.

Each of these areas and more are introduced in this handbook. The many examples help to show how the techniques improve manufacturing performance. These are the techniques that many multinationals are now using to compete around the world. The following is an overview of the contents of each chapter.

Chapter 1 introduces the concept of Operator Empowerment. It describes four different levels of empowerment that are appropriate to various manufacturing environments. The benefits of operator empowerment are explained along with various organisational structures that may enable the factory to empower in a controlled manner.

Chapter 2 describes Total Quality Management tools such as cause-and-effect diagrams and failure mode and effects analysis, and introduces statistical process control methods.

The chapter aims to demonstrate how these tools can be applied to reducing variation within the process and improving overall quality.

Chapter 3 explains the ISO 9000 and QS 9000 quality systems and describes the steps to be taken when considering registration. Quality systems are a feature of modern manufacturing, and accreditation is often a prerequisite for doing business.

Chapter 4 describes how to maximise the effectiveness of equipment through the use of techniques such as Total Productive Maintenance. The chapter describes how to utilise the technical resources available to the company in order to maximise equipment uptime and improve productivity.

Chapter 5 describes the Just-in-Time inventory management programme. The building blocks for JIT manufacturing are structured flow manufacturing, the kanban system and JIT deliveries. Each of these steps is described in detail.

Chapter 6 introduces Computer Simulation and describes how this technology helps with decision-making by enabling the user to experiment with various options before implementing change. The user can experiment in a virtual world created on the computer screen to analyse options and verify the expected effects of change.

Chapter 7 explains Process Reengineering and how it can benefit manufacturing. Process reengineering involves identifying the processes affecting performance; brainstorming to investigate how the process could be redesigned to achieve improved performance; and implementing the redesigned process. Each of these steps is fully explained in the chapter.

Chapter 8 describes Automation Technologies such as flexible manufacturing systems, and how they can contribute to manufacturing performance.

Chapter 9 introduces Project Management techniques. Techniques described include critical path analysis and resource optimisation, among other such useful tools. The ideas of Lewin on change management are also presented as an aid to the project manager in overcoming resistance to change and improving the project's chance of success.

Chapter 1

OPERATOR EMPOWERMENT

INTRODUCTION

Operator empowerment is a new approach to managing human resources on the production floor. It seeks to create an environment where operators and management can work as a team by attempting to reduce the barriers that can exist between them so that both sets of people work together.

Operator empowerment means formally involving all employees in achieving company targets. Everybody should be aware of what is expected of them and should have the means, the autonomy and the enthusiasm to request support to overcome obstacles to success. The extent of employee empowerment varies from company to company. In one company, it may simply involve shopfloor employees discussing their area's performance with managers or supervisors at regular intervals. At a more complex level, it may mean shopfloor employees forming empowered teams and taking responsibility for many of the tasks traditionally associated with the area supervisor, the maintenance department or the finance department.

For example, a team may take on responsibility for scheduling work through the area, or ordering its own raw materials. The level of operator empowerment suitable to a company depends on the complexities involved in achieving targets and the culture existing within the company. Theoretically, a group of shop floor operators could move from having regular meetings with management to gradually performing all the tasks currently done by the area supervisor. Generally speaking, any repetitive supervisory task could be considered suitable for moving into the realm of operator responsibility.

Factory employees, from technicians to machine operators, have a vast potential which can be harnessed for the benefit of the company and improved work environment. Consider, for example, the level of community work undertaken by many employees outside of work hours, raising thousands of pounds for charities, taking part in politics, holding club meetings and organising events. When these abilities are utilised within the work environment, the result is more job fulfilment for the employee and improved operating performance for the factory.

This changed relationship can be understood from a typical scenario that occurs when there is a production problem. On one occasion, I was responsible for maintaining equipment and was attempting to fix an electrochemical machine along with some other engineers. We were having difficulty and the work continued into the next shift. The second shift operator came to the machine and enquired if it would be down all night. I said it probably would, and the operator responded "good!". She saw an opportunity for a quiet night. She would go to the supervisor to be reallocated, but because it was unplanned downtime, it would take time for the supervisor to organise new work.

This story helps to explain the purpose and goals of operator empowerment. It can be considered as a programme aimed at changing this operator and other operators' attitudes. The operator perceived equipment downtime as somebody else's problem. Employee empowerment aims to redesign work and structures so that the operator has an interest in the machine's functioning and in the production process running well. Empowered operators perceive themselves as owners of the equipment, of the process and of the production area's performance. They can see the link between their immediate actions and the performance of the area in general and are recognised for their contribution.

DEFINING OPERATOR EMPOWERMENT

Operator empowerment is about changing the approach to decision-making within a company. The system pushes responsibility and authority for decision-making down the hierarchical chain to the operators who first come into contact with the problem. The result is a fast corrective reaction to problems, or a quick response to customer needs.

This approach differs greatly from traditional decision processes such as rule-based decision-making or the constant referral of decisions to one's supervisor. With rule-based decision-making, a procedure exists to define how any conceivable problem should be handled. In the Post Office, for example, procedures exist which describe how letters with illegible addresses or no stamp should be handled. For many companies, there are too many types of conceivable problems to write procedures which would cope with all. Most companies have procedures to cope with common problems, and exceptions to the rule are referred to management for decision-making. Management are responsible for day-to-day decisions such as production scheduling, responding to quality problems, organising training, analysing productivity levels and other typical issues. This system can work well in companies with low levels of change, but as the operations become more complex and more decision-making is required, the supervisory and managerial structure becomes overloaded. They are expected to respond to too many exceptions to the rules within too short a time-span. The result is a company that is slow to respond to its customers' needs.

The solution is to train the workforce so that these problems can be resolved by the people who first come into contact with them. In this way, machine operators and technicians become responsible for the day-to-day decision-making.

SUITABILITY

Employee empowerment can raise employees' participation levels at work, but not every company is interested in empowering their employees. Some companies deliberately avoid empowerment. They do not want to spend the time or the money needed to develop their workforce. They have analysed the potential of empowerment and decided it is not for them.

Many of these companies have made the right decision. The benefits of empowerment are much reduced when:

- The work environment is static;
- Equipment and production processes are simple and stable;
- Production is high volume and repetitive;
- Operators have low career aspirations, are low skilled and low paid;
- Managers tend towards an autocratic style.

In this sort of environment, managers consider there to be few opportunities for employees to improve factory performance. Rules and regulations are easily established to govern how to handle various situations. Why repeat the process of rediscovering solutions to old problems? Eliminate the problem or repeat the fix. Bowen and Lawler (1992) ask the question "what is the value-added from spending the additional dollars on employee selection, training, and retention necessary for empowerment?".

Conversely, companies operating in a fast-changing, complex work environment have more to gain from empowerment. Empowerment benefits companies when:

- The work environment is dynamic;
- Equipment and production processes are modern and complex;
- Production is low-to-medium volume for a diversified product range;
- Operators are highly skilled and have high career and social aspirations;
- Managers tend to have a participative management style.

The unpredictable environment created by these factors makes employee empowerment attractive. Management have to cope with a higher quantity of problems. Employee empowerment allows them to delegate responsibility for problem-solving to production personnel. In so doing, it makes the personnel own-

Operator Empowerment

ers of their decisions and encourages fast and fle[x]
making by the people closest to the problem. It c[reates an]
environment where the chances of success are mu[ch...]

Management should consider where their wor[k]
is positioned with respect to these two extremes. It may be that certain departments appear to have the characteristics of a static work environment while others are more dynamic. For example, in a high-volume production plant, operators may have little opportunity to contribute, while maintenance technicians or customer service departments may benefit more from empowerment. A proper analysis of company characteristics can save time, money and effort.

IMPLEMENTATION GUIDELINES

It is not possible to provide a universally applicable implementation plan. Companies are like people in that they have different characteristics and attitudes. However, a guide can be provided which, at a minimum, provides a framework against which the reader can compare their own plans. The guide is divided into three phases:

- Preliminary work;
- Analysis work;
- Implementation recommendations.

These three phases provide a framework which takes the project from beginning to end.

Preliminary work involves conducting an employee survey, establishing company goals and performing a cost-benefit analysis. During the preliminary phase, the company establishes the suitability of empowerment. This involves taking a close look at the needs of employees and attempting to match them to the needs of the business. Goals are set and the cost of achieving them through empowerment is evaluated. When this phase is complete, the project team should be confident that empowerment is the correct solution for their organisation. It should identify the extent to which empowerment satisfies company goals and enables the team to progress to phase 2 — analysis work.

Analysis work involves determining the level of empowerment that suits the company's requirements. It also means identifying training needs and designing the structures that enable employees to take on their new roles. Examples are given of structures adopted by some companies which may help the project team design structures suitable to their company.

Implementation recommendations provide advice on implementation and suggest ways of avoiding common pitfalls.

Creating an empowered workforce can be difficult, but the framework described above should be of assistance. When the initial groundwork indicates that empowerment is the way to go, then the rewards for the company can be very significant.

Getting Started: Establishing the Team

The first step in any project of this size is to establish a team to drive the project. Company-wide empowerment needs top management support and therefore it is expected that top management will actively participate in the project team.

Company circumstances vary, but in my experience of employee empowerment programmes, teams tend to have strong human resource participation and generally a project champion who reports directly to top management. In one company, a top management position was created for the person to run the empowerment programme. Other members of the team may be industrial engineers, union representatives or departmental heads.

Consultants also appear to be widely used, but to varying degrees. One company hired consultants to train facilitators who then trained employees. Another company hired consultants to play a much more active role in implementation. These consultants trained employees while introducing new work structures. Both methods work well, depending on the situation within the company.

Members' participation levels also vary but typically, the project champion works on the programme full time. Other team members' participation levels vary as the project progresses.

Starting the Project: Phase One — Preliminary Work

Preliminary work consists of three stages, namely: conducting an employee survey; establishing company goals; and performing a cost-benefit analysis.

Employee Survey

The aim of the employee survey is to determine the concerns people have regarding their work. It aims to gather information which reveals how people perceive their work environment and how they consider it could be improved.

The survey must indicate if the company has the characteristics that make empowerment attractive, as discussed earlier. Answers are needed to the following types of question: Do operators, technicians, engineers and managers perceive the company to be static or dynamic? Is there a high degree of skill required to operate equipment? Is there a high level of uncertainty in how equipment and processes will perform from one day to the next? Do employees have the career aspirations to take on an empowered role?

There are two common ways of gathering this information. One way is to design an employee questionnaire which asks employees and operators the necessary questions.

A second way is to hold "skip level" meetings. Skip level meetings are informal, separate meetings between top managers and operators, technicians and engineers. The idea is to ensure that the members attending any meeting are from the same level within the organisational structure and — most essentially — to ensure that none of the attendees' direct managers are present. In this way, conversation should be less inhibited and a more realistic picture presented of what is actually happening within the workplace.

Skip level meetings can be very informative. I worked in a company that was very keen on skip level meetings. The plant manager learned some very useful information at these meetings. He was delighted to discover that the operators wanted to play a more active part in equipment maintenance. They felt that many of the fixes were repetitive and didn't like having to call maintenance. They disliked having to wait around for a technician to be freed up and then having to wait for the repair. Some operators preferred the machine to run continuously because it benefited the company. Others liked continuously running machines because it made their time at work appear to go faster! This information encouraged the plant manager to train operators to perform repetitive fixes and some preventative maintenance work on equipment.

Establishing Company Goals

The concerns and aspirations of employees have to be reconciled with the objectives of the company. The aim is to create a win–win situation where the company can improve performance while addressing issues raised by the survey. The results of surveys will be particular to the traits of the individual company. One company that used an employee questionnaire to gather the necessary information found that employees were not motivated, possessed no sense of team work and received little job satisfaction. They indicated that they wanted extra responsibility and ownership for day-to-day decision-making. This was certainly fertile ground for introducing the concepts of empowerment!

The company's management also had objectives. They wanted to flatten the organisational structure, thereby pushing responsibility for decision-making down through the organisational hierarchy. Another objective was to simplify pay grades and make them more relevant to the needs of the company. These two positions provided ample opportunity to find common ground.

Companies will vary in what they want to achieve through empowerment programmes. Some objectives may include:

- Reduced voluntary employee turnover;
- Reduced absenteeism;
- Improved quality and productivity levels;
- A flexible, skilled workforce.

Whether the company objective is reduced absenteeism or improved productivity, targets should be set and the cost-benefits estimated. Empowerment costs money and should therefore be scrutinised in a cost-benefit analysis.

Cost-Benefit Analysis

The major initial costs for empowerment are consultancy, training and, very often, a reduction in productivity as employees become accustomed to their new roles. Ongoing costs can be incurred when wage structures are redesigned to encourage participation in the empowerment programme. Employees may want wage increases for taking on the extra responsibilities.

Some of the financial impacts of benefits may have to be guesstimated. It can be difficult to judge accurately the cost implications of high absenteeism, or high voluntary employee turnover. Guesstimates should provide a basis to enable the accounts department to determine the financial feasibility of empowerment.

Phase Two: Analysis Work

The preliminary work enables the programme team to have clear goals and an understanding of the cost-benefit relationship. The team has to decide the level of empowerment which will enable them to achieve their objectives. Four levels are described below, varying from a situation of no empowerment in an autocratic environment to full empowerment where employees are actively encouraged to develop themselves. Most companies find a level suitable to their needs somewhere in between these extremes.

Level one is synonymous with no employee empowerment. There is autocratic decision-making where top management enforce their decisions throughout the hierarchical organisational structure. There is little or no consultation with their subordinates in arriving at decisions and management exercise tight control over the actions of their subordinates.

There are advantages to this approach. It tends to suit companies that are resigned to coping with high employee turnover and where new employees often need strong direction. It also suits companies that have well-defined rules and regulations. Decisions are generally quickly implemented because there is seldom opportunity for discussion.

The disadvantages relate to the lack of utilisation of employee potential and the low employee motivation generated by this approach. Absenteeism may be high as employees do not see a future for themselves within the structures of the company and find their work repetitive and banal. They do not perceive the value of their work in contributing to company success.

Level two is the minimum level of two-way interaction between management and operators. It involves putting some structures in place which provide operators with an opportunity to communicate their ideas to management. This may take the form of an employee suggestion scheme or regular all-employee meetings. Management decide whether or not to accept and implement the ideas. With suggestion schemes, the accepted ideas are rewarded. Level two empowerment is easy and relatively cheap to implement and does not distract the employee from performing their normal daily workload.

A suggestion scheme has a number of advantages. Managers gain access to potentially good ideas at relatively low cost and the scheme provides an opportunity for employees to make some extra money and to participate within the company.

This level has similar disadvantages to the previous level. It doesn't cater for employees with career aspirations and it only makes a minor contribution to employee motivation.

Level three aims to increase directly the operator's contribution to decision-making and increase their ownership of the process. It achieves these aims by redesigning their work structure. Regular opportunities are provided during the working day or week for the operators to contribute their opinion, make recommendations and report progress.

In an autocratic work environment, top management go to middle managers to find out what the problems on the line are, what has caused these problems and what they are doing to resolve them. Middle managers get their information from a variety of sources, but they seldom get it by asking the operators. Within a level-three empowered organisation, middle managers, supervisors and engineers ask similar questions of the operators. The structures are in place which allow the operator to gather information and to analyse and present it. Ideally, there is a change of responsibilities between management and operators. Operators take on responsibility for resolving problems which were previously perceived as management problems. Management take on new roles in facilitating operators to find solutions to production problems. Decision-making on a range of issues becomes the responsibility of operators.

There are a number of structures that achieve this goal. One structure might involve middle managers pulling together their team of machine operators on a weekly basis to review progress. An open atmosphere is encouraged where operators are not inhibited to speak up. Problems such as production falling short of targets are raised. The reasons may be varied — for example, a high level of damaged raw materials resulting in machines running out of good material to work on. The group may nominate a person to keep samples of the damaged raw materials and to contact incoming quality control for a problem prevention plan. The person will report back at the next weekly meeting. This meeting provides an opportunity for the middle manager to relay information to operators. They can report and update on the extent of which the monthly target has been achieved and what problems are anticipated. Maybe there is an upcoming customer audit, and everybody must be extra vigilant in keeping their machines clean. Holding regular meetings to discuss problems creates a team atmosphere on the production floor, and makes operators co-owners for decisions taken.

Another structure may involve meeting on a daily basis and delegating more authority to operators. In this structure, the team meets each morning to discuss the previous day's performance and to distribute work amongst its members. Perhaps the supervisor sets up a rota such that each operator has an opportunity to be charge-hand for a week at a time. The operator/charge-hand organises and chairs the daily meeting, gathers relevant information and generally performs the tasks associated with typical supervisors. The supervisor has more time for other, more beneficial work, yet supports the chargehand throughout the week. The net result is that all operators experience life from the other side and gain a greater appreciation of how they can contribute to success.

Level three empowerment involves management asking operators for help in managing and in resolving issues. Over time, a sense of ownership tends to develop as operators recognise that they have a voice and that their contributions are welcome.

Level four empowerment gives even more autonomy to operators. They become responsible for practically all the tasks associated with supervisors and are expected to run the production area amongst themselves. This may involve production scheduling, allocation of operators between machines, resolution of quality and production problems, among other responsibilities. Some level four teams, for example, even partake in the selection of new employees or the preparation of budgets for their area.

A structure that supports level four empowerment involves handing over responsibility for the production area to the team of operators. The operators nominate a colleague from within their ranks to lead them for, say, a three- to six-month period. A rota of suitable leaders can be drawn up. The chosen person performs the tasks typically associated with the supervisor, but receives support throughout from their colleagues and management through daily meetings. The leader gathers information and reports to managers. Level four empowerment demands a lot from its participants. They must have the educational abilities and personnel skills typically associated with supervisors. Level four empowerment results in a self-managed team that can cope with a high level of decision-making. They are empowered to respond to problems as they occur.

There are many other structures that enable a company to empower its staff to various levels. The ideal configuration, the intervals between meetings, the distribution of responsibilities, are all compromises between the abilities and aspirations of operators and management.

Matching Work Environment to Empowerment Level

Each of the structures described in levels three and four empowerment places extra responsibilities on operators. It becomes very important that the employee survey indicates the willingness of operators to take on these new roles and that management perceives the need for the new structures. Otherwise a confused mismatch develops. Figure 1.1 below emphasises the links between the work environment and the most suitable empowerment level.

FIGURE 1.1: OPERATOR EMPOWERMENT AND ENVIRONMENTAL FACTORS

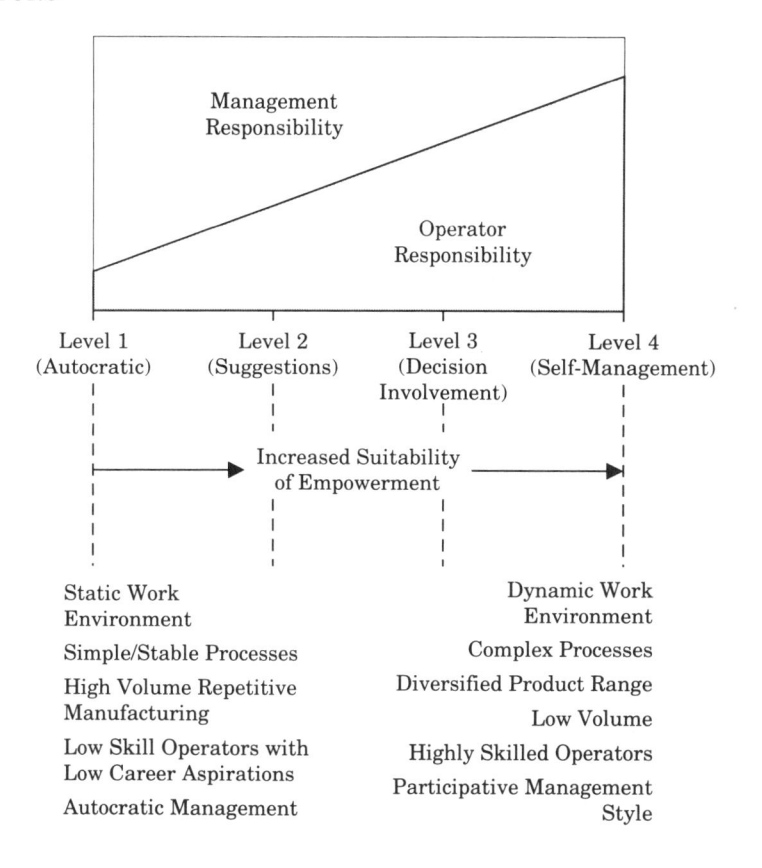

Level 1 (Autocratic)	Level 2 (Suggestions)	Level 3 (Decision Involvement)	Level 4 (Self-Management)
Static Work Environment			Dynamic Work Environment
Simple/Stable Processes			Complex Processes
High Volume Repetitive Manufacturing			Diversified Product Range
			Low Volume
Low Skill Operators with Low Career Aspirations			Highly Skilled Operators
Autocratic Management			Participative Management Style

Increased Suitability of Empowerment →

Management Responsibility / Operator Responsibility

What do Operators Gain from Empowerment?

Empowerment systems are designed to enable operators who, through their personality traits and abilities, are able and willing to take more control of their work environment. The level of empowerment that is practicable is limited by their abilities and willingness.

According to Taylor's ideas on factory efficiency, operators were given a narrowly focused job and became highly specialised and highly efficient at performing tasks. This concept may have been suitable in an era where second-level education was uncommon and factory operators needed a lot of direction. In today's workplace, operators are more educated when they enter the workforce, and many are no longer satisfied to perform minimalistic tasks that require no intellectual work. Herzberg's studies into motivational theory confirm the desire of operators to have more responsibility in their work.

Herzberg and his colleagues conducted a survey of company employees and asked them to recall times when they felt exceptionally good about their work. The purpose of the survey was to identify factors affecting job satisfaction. The interviewer asked questions about events leading up to the occasion of job satisfaction and classified the reasons under a series of headings. Herzberg repeated the questioning exercise for periods of job dissatisfaction. Again, the responses were classified under various headings.

Herzberg found that the events leading to job satisfaction are very different to those leading to job dissatisfaction. From Figure 1.2 below, there are five strong criteria that lead to job satisfaction. However, lack of these factors was not found to be a significant cause of job dissatisfaction; rather, causes of dissatisfaction are poor salary, unstructured company policy and administration, lack of recognition, poor working conditions, and poor supervision. Herzberg concluded that the two feelings are not the opposite of each other, and are driven by different sets of needs. Herzberg called the causes of dissatisfaction "maintenance" or "hygiene" factors and the causes of job satisfaction "satisfiers" or "motivators". The thickness of each of the lines expresses how long the effects of each factor lasts. For example, Figure 1.2 shows that job satisfaction resulting from having responsibility lasts longer than satisfaction associated with achievement.

FIGURE 1.2: ELEMENTS OF JOB SATISFACTION/DISSATISFACTION

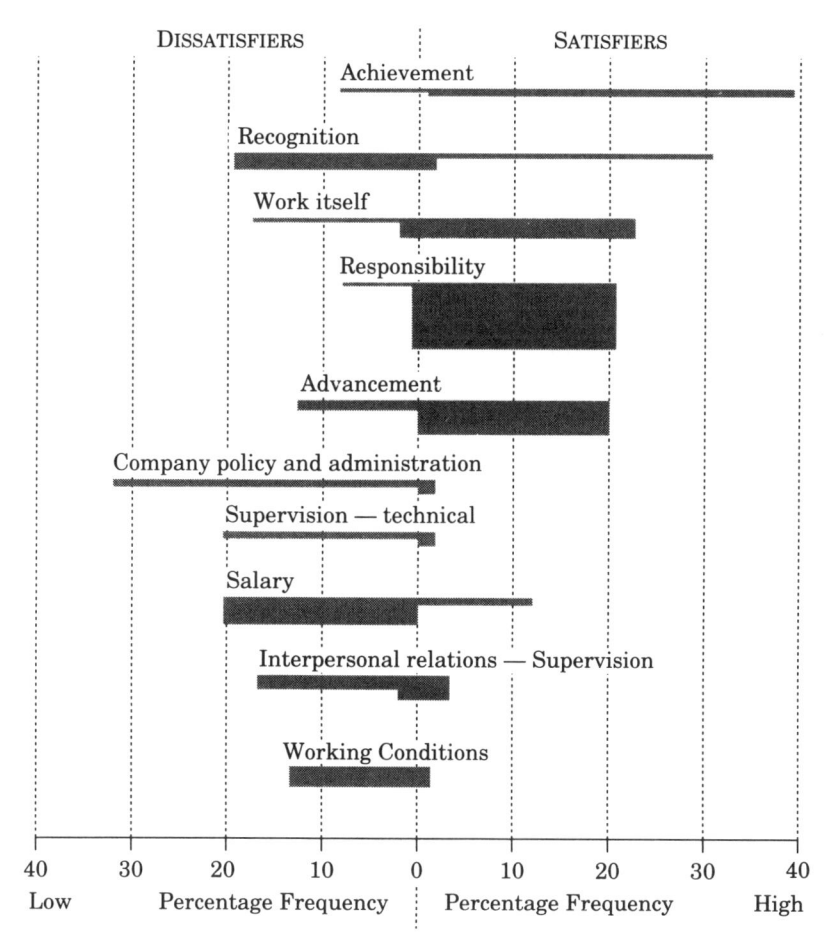

DISSATISFIERS	SATISFIERS

Achievement
Recognition
Work itself
Responsibility
Advancement
Company policy and administration
Supervision — technical
Salary
Interpersonal relations — Supervision
Working Conditions

40 — 30 — 20 — 10 — 0 — 10 — 20 — 30 — 40
Low — Percentage Frequency — Percentage Frequency — High

Herzberg related the two sets of needs to two biblical characters, Adam and Abraham. According to this analogy, there is an Adam and Abraham in each of us. Adam experienced a rude awakening when he was banished from the Garden of Eden and forced to survive in the world. His immediate needs were for security, safety, a sense of belonging, clothes, and food, all essential for survival. These needs are related to our "maintenance need"; the lack of any one of them results in dissatisfaction. In the context of the work environment, the employee has survival needs that must be appeased. These include the need to earn a suitable salary, the need for job security and the need to belong to a group. Job dissatisfaction is avoided by resolving problems in these areas through improving company policy and supervisors' interpersonal skills, among others. However, Herzberg points out that no level of improvement in these factors will result in job satisfaction. The second set of factors, "satisfiers" or "motivators", provide job satisfaction and are associated with Abraham.

Abraham experienced personal fulfilment through his relationship with God and the many accomplishments he achieved in his lifetime. His life is characterised by self-development, growth, learning, and achievement. Employees also have these needs and fulfilling them results in job satisfaction. Motivation of people involves giving them responsibility, opportunities for achievement, and recognition for work well done. They must be encouraged to develop and given the opportunity for advancement when vacancies arise. Herzberg recommended job redesign to incorporate these characteristics into job designs.

He promoted the need for job enrichment through expanding the opportunities for employee development within the context of the job. This does not mean giving additional mundane tasks to employees to relieve the tedium of their work. Rather, it involves incorporating decision-making responsibility in the employee's job. The recommended approach is to identify ways of removing some controls governing the job and making the employee accountable for performance, granting them the authority and freedom to take actions required to improve performance. These ideas are in line with the concepts of level three (decision involvement) and level four empowerment (self-managed).

Herzberg tells us that people have a desire to take control of their work environment and experience job satisfaction as a result. It is this improved satisfaction that leads to reduced absenteeism, higher morale, reduced voluntary turnover and therefore improved overall performance.

Designing Jobs for Empowerment

Job design is highly dependent on the circumstances prevailing within the manufacturing facility. Different manufacturing processes place different requirements on operators and these requirements have to be factored into the job design in such a way as to promote job satisfaction. Hackman and Oldman (1976) describe important elements of jobs that support Herzberg's theory of motivation. These are:

- Skills Variety
- Task Identity
- Task Significance
- Autonomy
- Feedback.

Any job can be analysed to evaluate the role played by each of these factors. If we consider people who appear motivated, especially operators, and analyse their job content, we generally find that these five elements are prominent. The complete absence of any factor reduces the motivational and job satisfaction levels. Over time, I have worked with many motivated groups, but one in particular stands out. Their motivational levels could be gauged by:

- The level of improvement suggestions forthcoming;
- The high standard of training provided for new operators;
- The willingness to request assistance when needed and highlight problems;
- The help provided to technicians in problem-solving and equipment maintenance.

The particular group I mention were responsible for equipment which was over 15 years old and was manufactured by a company on the other side of the globe. There was no technical support available from the original equipment manufacturer and spares were generally difficult to acquire. There was a core group of people operating these machines for over 5 to 7 years, and as a result they had gained much experience over that time. They took a personal interest in the performance of the equipment. When their work is analysed under Hackman and Oldman's criteria, it is seen that each of the five factors is present, to some degree, in their jobs.

Skills Variety

This feature refers to the range of skills that each worker has to use in performing their job. The need for a variety of skills implies that the operator must have scope to learn continuously within the context of the job. Over time, they develop a unique set of talents that enables them to make an individual contribution to their work This emphasis on skill variety increases the sense of achievement for work well done.

In our example, the age of the machines meant that sustaining their throughput levels required quick reactions from operators, using technical skills to make critical repairs. Over time, their technical competence on the equipment improved, and as a result they were a valuable resource for all technical staff in resolving equipment and quality issues.

Some jobs — highly repetitive monotonous tasks in particular — do not facilitate skill variety. There is no opportunity for operators to utilise a range of skills. It is difficult to make a monotonous job interesting. Some companies use job rotation to relieve tedium, others invest in automation technologies to perform these tasks. There is no ideal solution, since job rotation is temporary and automation can be expensive.

Task Identity

This means performing a job that has a clearly identifiable beginning and end, with a visible outcome. A computer games manufacturer organises production so that each game system is produced by approximately five or six people. Each person performs a series of tasks that leads to the completion of each unit. In each of the jobs, there is a sense of achievement as components are transformed into computer game systems.

In the earlier example of operators using old equipment, they perceived their task as two-dimensional. They were responsible for maintaining throughput and also for sustaining their equipment. These two levels provided the task identity for the group.

Task identity is generally low in situations where the work cycle repeats every few seconds or minutes, as in the case of manual inspection tasks, where the task can be monotonous and repetitive. In some situations, the task identity can be increased by increasing the cycle time for the task, thereby adding a greater sense of fulfilment and achievement. In the case of manual inspection, the operator, for example, may be made responsible for evaluating the causes of defects in addition to performing on-line inspection. The discovery of a defect enables the operator to perform a follow-up analysis, which expands the role of inspection. The operator now has a failure analysis role, as well as being an inspector; the overall task identity has improved.

Task Significance

Task significance refers to the degree to which the task has a substantial impact on the work of other people. Some jobs have a readily identifiable high task significance, but for others, it is less obvious. It is important to be able to identify the task significance of all jobs, especially those that appear to be minor tasks where the worker needs to be motivated, perhaps even more so than usual, by having their task significance identified and recognised. For example, when a customer threatens to withdraw their business because of the high level of rejects being received, the person performing the manual inspection task suddenly receives a lot of managerial attention. Often, manual inspection can be perceived as having a low task significance until, of course, the ill-effects of ignoring it are realised, as in the case of customer complaints or threatened business. However, it is important not to wait until a problem arises before making the person responsible for the inspection task aware of the significance of their job. In one factory, there were two to three operators who were particularly good at manual inspection. Their dexterity at manipulating the part under inspection was phenomenal to observe. They were much more alert than other inspectors. What is important to note here is that they were made aware of the task significance and they were recognised for their ability and individual contribution. Although the task had low levels of task identity, it had high task significance, especially if a customer returned product. In our example, task significance was very high, since the machines were troublesome and the support of the operator in troubleshooting was critical. Task significance enables people to take pride in knowing that their work is contributing to the overall organisational goals.

Autonomy

Autonomy is the degree to which the operator has freedom to make decisions regarding their work. In situations where the operators merely follow their supervisor's instructions, the level of accountability required from the operators is low. People only feel responsible for the outcome of a particular task when they have had input into the design and co-ordination of that task. Autonomy in decision-making increases the level of responsibility delegated to the operator and is therefore a major source of job satisfaction. In our example, the equipment operators were always involved in decision-making. When technicians wanted to make improvements, they always evaluated the impact of the modification with the operators. The operators were the first line of defence for quality and equipment. They were relatively autonomous in how they performed their duties; their supervisors trusted them to take the appropriate action and contact them if further problems arose. Their level of autonomy was synonymous with level three, described earlier.

Feedback

Feedback is information about performance which should allow individuals to review their performance and others to view it. If their behaviour is positive, this must be communicated to them and reinforced. If their behaviour is negative, the feedback should allow the individual to alter performance accordingly.

It is critical that feedback should be given in the correct manner. Negative, or indeed positive feedback, if expressed in the wrong way, can be very demotivating and damaging. When giving feedback it is important to analyse the behaviour, not the personality, concentrating on specifics and facts, not common beliefs and judgements. Most importantly, it should highlight the consequences and should be supportive, not threatening, in either reinforcing correct behaviour or altering more negative behaviour.

Everybody likes to know how they are performing, and especially to have their achievements recognised. Feedback should be built into the job design so that people can measure their own performance through tasks such as quality checks. Regular self-inspection helps to heighten awareness of poor performance. Feedback also comes from supervisors and managers who monitor performance and update operators with regard to how their contribution is perceived. Supervisors should *always* ensure that feedback is a regular feature of work, not only when things go wrong. Feedback should also indicate to operators that management are aware of the work being done and that they appreciate that everything is running smoothly. It is unfortunate that, sometimes, the only way to gain recognition is to solve a crisis rather than to prevent one occurring.

In manufacturing environments, feedback often comes from workstation operators, who have to use the output from previous operations. Any quality problems or material shortages are rapidly highlighted for attention. Feedback encourages good operators and controls poorly performing operators. Good operators gain a sense of achievement from work well done, while less attentive operators are aware that performance controls are in place.

In our example, the equipment operators received feedback from supervisors and technicians on a daily basis. Their throughput levels also provided a measure of their performance, and there were often contests to break the overall highest throughput level. They operated the machines that were acting as a bottleneck on the line, and thus most restricting on performance, so that any variation in performance was quickly investigated. Exceptionally good performances were congratulated and poor performances were further investigated.

Each of these factors — skill variety, task identity, task significance, autonomy and feedback — can be linked back to Herzberg's theory of motivation. When these factors are designed into the job, there is higher scope for successful operator self-growth and empowerment. These are the elements that generate a sense of responsibility and achievement, which results in a high level of motivation.

Training

The level of empowerment delegated to operators has a major impact on the level of training needed. Level two empowerment will need little training other than to introduce the concept and mechanisms of the suggestion programme. However, levels three and four make significant demands on operators. The training programme will have to educate operators so that they are able to perform their new responsibilities. The person designing the training programme must identify:

1. What the employees need to know
2. What performance standard is considered acceptable
3. How performance is measured
4. Who carries out the training
5. If re-certification is required
6. How the re-certification process will operate
7. What the attitude of the people involved will be.

The level of training required depends on the responsibilities of the operators. The following table shows a few examples:

Possible responsibilities	*Possible new training needed*
Reporting production numbers	Computers and graphing
Reporting on quality	Statistical Process Control
Resolving problems	Problem-solving methods
Organising the workload	Interpersonal skills
etc.	etc.

Management will also need training to enable them to adapt to their new environment. They will need to understand how they can achieve their goals through the empowerment system and how they can contribute to the success of the system.

In one example, a manufacturer arranged company-wide training for all employees. They received four days training on empowerment at the beginning of the project. During the training session, they learned about the company's competitive position and where the sources of immediate competitive threats lay. The concept of empowerment and its importance was explained. Operators learned how they could contribute more effectively to their company's success. They were then introduced to techniques for problem-solving and these techniques were tested using practical examples. Supervisors, engineers and managers received extra training on how empowered operators could improve the work environment. Employees returned to their work environment with a basic knowledge of empowerment and problem-solving techniques.

Another approach to training is to introduce empowerment on a pilot basis and for the training instructor to evaluate and satisfy the group's training needs as they occur. This approach ensures a high correlation between the eventual training course and the needs of employees.

Phase Three: Implementation Recommendations and Pitfalls

The successful matching of company goals, the work environment and employee aspirations enhances the opportunity for the introduction of a suitable level of empowerment. Analysis of the current situation should establish an empowerment level and structure suitable to company goals while improving employee morale. Ideally, a win–win situation encourages top management and employee support for the project.

Level two empowerment is a low risk option. It is relatively easy to set up and does not have any adverse affects on productivity during its installation. There is no emphasis on operator training courses which temporarily redirect operators away from their machines. Very often these suggestion schemes receive a flood of suggestions, especially during their early-life phase. It is important that there are structures and well-defined criteria in place to enable management to evaluate and judge each suggestion fairly. Typical issues associated with the implementation of a suggestion scheme are:

1. How shall ideas be evaluated? A cross-functional, impartial team usually provides a forum where suggestions are evaluated as to which recognition category it falls into, e.g. non-runner, good safety idea, good productivity idea, etc.

2. How shall the recognition system operate? A tiered award system can direct people's "suggestion thinking" process to safety or productivity issues.

3. How will suggestions be implemented? Are there financial constraints limiting which suggestions can be implemented?

Suggestion schemes tend to fail when operators have to wait too long to get a response to their idea, or when they don't understand the reasoning behind a rejected idea.

Empowerment levels three and four are high risk options. It is easy to misunderstand the needs of either the employees or the company and thereby waste time, management credibility and money. For these reasons, a pilot implementation is recommended.

Levels three and four may involve negotiations with a union on pay structures and it generally means investment in training and consultants. A pilot implementation helps to ensure that the benefits to the company exceed the costs involved. It allows the company to experiment and to withdraw the project if results are unfavourable.

The pilot group will need clear, unambiguous instruction as to what is required of them. They will need to know the objectives of the exercise, what information they are expected to gather for the meetings and how to gather and interpret the information. Relevant cross-functional personnel may be expected to attend and support these meetings — for example, equipment maintenance, process engineering or quality staff.

In one company, operators were empowered on a pilot line. The shifts were designed to overlap and a meeting was held at the beginning and end of each shift to discuss performance. The operators gradually assumed responsibility for the different supervisory tasks and made many improvements to their area. There was no need to sell the concept of empowerment to the other factory employees, who ended up asking management to empower them, because they knew they could be just as good as the pilot team.

Level four empowerment is very dependent on the capabilities of operators. Operators will need to have strong career aspirations and good supervisory skills in order to take on the extra responsibilities described by level four empowerment. When Toyota was hiring staff for its greenfield site in America, they chose people who had previously held responsible jobs, such as ex-police officers, farm managers and entrepreneurs. They looked for people who could take on responsibility and leadership. One company hired technicians to operate their equipment but then delegated a broad range of extra responsibilities to them. These people provided fertile ground for level four empowerment. They had the capability to absorb the role of the supervisor into their workload and thereby manage themselves.

What Happens to Supervisors?

In organisations dedicated to levels three and four empowerment, the need for direct supervision of operators is much reduced, or ideally eliminated. In many empowered work environments, operator teams report directly to the middle manager at the level directly above that of the supervisor. Indeed, a company objective for undertaking self-managed teams may be to reduce the number of hierarchical levels within the organisation and therefore eliminate the need for supervisors. Supervisors lose control, authority and potentially status within the organisation.

In spite of the potentially negative effects on the role of the supervisor, their support of the project is vital for its success. In level four empowerment (self-management), they will have to train the team to adapt to their new responsibilities and show them how to get things done within the organisational framework. During the implementation process, they will have to hand over their responsibilities gradually as the team matures and its members develop their capabilities.

Empowerment can be an exercise in reducing headcount, or it can be an opportunity to free up people who can contribute in new ways to the organisation. Supervisors and front-line managers have a wealth of knowledge and expertise built up over years that can be utilised in new and more profitable ways. The implementation team needs to consider who will be affected by the changes being introduced and how these people can be helped to make the transition. It is critical that the team creates opportunities, at the other side of this transition, for supervisors and other front-line managers. Without their support, successful implementation will be very difficult, and the organisation could loose valuable resources.

In many cases, the role of the supervisor changes to that of a facilitator for the team, helping its members become accustomed to their new roles and working with them to troubleshoot problems as they arise. Since the facilitator is not directly responsible for team members and their performance, they have more time available to expand their role into new areas rather than concentrate on previous repetitive tasks, and can offer facilitator services to several teams.

Some companies utilise the expertise of supervisors by creating new improvement project roles for them within the company. They take on the role and responsibilities previously associated with project managers. In this way, the company gains an expert team, fully familiar with the organisational system and capable of accelerating the pace of improvement work.

Employee empowerment has a major impact on the traditional role of the supervisor. The company can view the introduction of empowerment as a means of better utilising the supervisors' capabilities or simply as a means of reducing headcount. The probability of success will be much improved when there is a clear, well-defined role for all employees after the transition process.

Pitfalls to Successful Implementation

The road to successful empowerment of operators is seldom straight. In many cases, managers abandon the project and are eager to return to the traditional practice so common in manufacturing. Their initial idealism is exhausted in the effort to make empowerment part of the organisational culture. There are many reasons for failure, some of which are listed here.

A poor understanding of empowerment can lead to attempted implementations where the philosophy does not suit the needs of manufacturing. Many companies survive by having operators perform standard jobs according to set procedures. Empowerment means chaos in these workplaces.

Another cause of failure is the difficulty of transferring the principles learned in the classroom onto the manufacturing floor. The culture of the operators and management may be completely opposed to ideas of empowerment. Unions may make the support of its members conditional on excessive demands. They may perceive these new systems only as ways of extracting higher productivity levels. Management may not be willing to share their authority with subordinates.

Empowerment is seldom an overnight success, and unfortunately, productivity levels often decrease at the outset of a high-level empowerment programme, often because employees are attending training courses or meetings. The initial drop in productivity may lead to management justifying the scrapping of the project; or if productivity is slow to improve, management may take back decision-making responsibility that had been delegated.

A pilot implementation helps to minimise exposure to these risks. The trial should demonstrate the improvements that can be achieved and should generate acceptance for the new system with the union. Management should ideally perceive the new system as freeing them up to undertake more long-term projects.

A programme of employee empowerment should be considered as a long-term business strategy. Once begun, it requires patience and commitment from everybody involved. Employees in various positions within the company will require training to understand their new roles and new interaction with work colleagues. It will require a dedicated project champion and a lot of senior managerial support, especially during its early-life stage. Once begun, there should be a commitment at top management level to support the process to its fulfilment. Having asked operators for their opinion and support, management and supervisors should actively listen and respond to the recommendations. When decision-making autonomy has been given to operators, they should be supported as they learn how to adapt to their new responsibilities. The well-designed training programme will provide individuals with the tools they will need in order to cope with their new responsibilities.

CONCLUSIONS

Employee empowerment means autonomy for the employee to carry out their work with the minimum level of supervision. The acceptable "minimum level of supervision" varies with the company's circumstances, the management's preferred working style and the capabilities of the employees. Each of these factors affects the suitability of empowerment for the company.

Four levels of empowerment have been described. At level one, autocratic supervision of employees predominates. This direct supervision style is sometimes necessary to ensure that decisions are taken quickly and to minimise time required to perform instructions. Autocratic supervision discourages employee participation in decision-making and can lead to frustration and hostility from employees who feel their attempts to contribute are stifled. Level two — suggestion involvement — means setting up a formal system which enables employees to make suggestions to management. The suggestions generally relate to recommended improvements on the production floor. A reward system encourages employees to contribute their ideas. Level three — decision involvement — actively encourages employees to partake in decision-making regarding production line improvements. Employees are encouraged to join task teams set up to resolve production problems. Level four is high involvement, where employees take responsibility for achieving targets and are given the autonomy to decide how they will achieve them.

The level of autonomy most suitable to a company depends on its particular circumstances, such as its preferred managerial style, the capabilities of its employees and the urgency with which decisions need to be made, among other factors. Training of employees and management is part of the empowerment process. Managers must understand how their roles change and employees must develop new skills such as problem-solving, how to hold meetings and how to analyse data.

Companies that adopt a suitable level of employee empowerment benefit from having a workforce that is proactive in contributing to the welfare of the company. Effective utilisation of human resources is a key factor for succeeding in today's competitive environment.

Chapter 2

TOTAL QUALITY MANAGEMENT TOOLS

INTRODUCTION

Quality was the battleground on which multinationals won and lost market share in the 1980s. In that decade, market share on a number of products went east to Asian companies. Companies like Toyota, Nissan and Honda expanded into the American car market. Korean companies such as Lucky Goldstar and Samsung became giants in the semiconductor market. Table 2.1 shows the extent of the transfer in market share.

TABLE 2.1: WORLD MARKET SHARE HELD BY US COMPANIES

Industry	1978	1988
Cars	29%	18%
Floppy Disks	66%	4%
Drams	73%	17%

Source: *Fortune* Magazine, 1 January 1990.

The good thing about the 1980s is that the level of interest in quality during that decade has resulted in quality techniques that are well-documented and refined. Companies are using these techniques to achieve defect rates as low as 3.4 parts defect per million. That means that 99.999 per cent of products are defect-free — in other words, almost zero defects. This is the goal for the year 2000 set by many top multinationals for all their subsidiaries. This quality level benefits companies in two ways. Firstly, they are producing less scrap, so their standard unit cost is lower, and secondly, since their product has a guaranteed high quality level, they can command a premium price on the market.

W. Edwards Deming is the guru behind many quality approaches being adopted by companies today. A key part of his philosophy is the elimination of variation. This concept is the foundation on which most successful quality programmes are built. Consider a beefburger manufacturer whose customers complain about variation in flavour or bone content being found in their burgers. The manufacturer is perplexed because he has, from time to time, personally taken the raw material that goes into his burger through the process and produced an excellent end product. The excellent end product defines his capability. It is the inconsistencies or variation in the process over time that prevents the same high standards always being achieved. The goal is to eliminate these variations from the ideal process.

There are two ways of ensuring product quality: a factory can concentrate on testing out all reject goods at the end of the production line; or they can build quality systems that concentrate on ensuring the product is initially produced correctly. Figure 2.1 below illustrates the two systems.

FIGURE 2.1: APPROACH TO QUALITY CONTROL

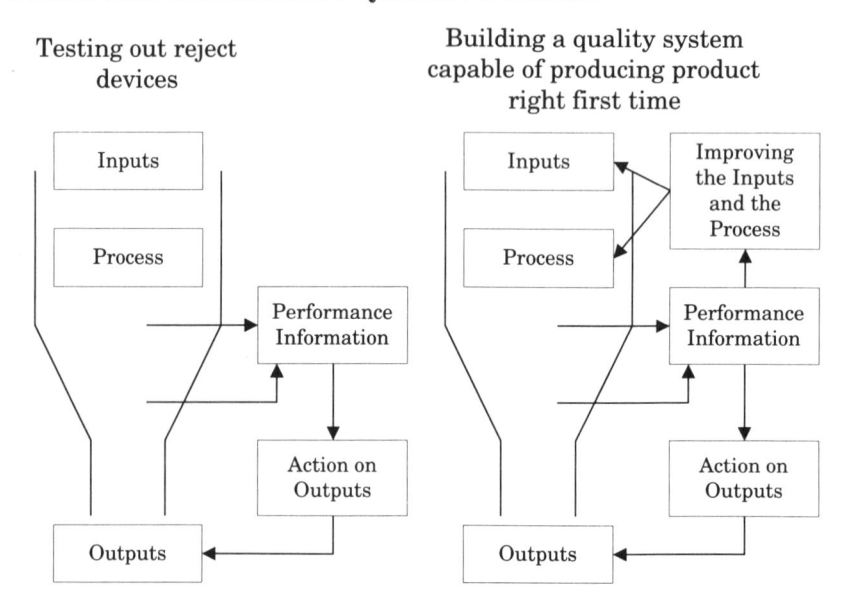

Testing out reject devices

```
Inputs
  |
Process
  |
  v
Performance Information
  |
  v
Action on Outputs
  |
  v
Outputs
```

Building a quality system capable of producing product right first time

```
Inputs ──> Improving the Inputs and the Process
  |             ^
Process         |
  |             |
  v             |
Performance Information
  |
  v
Action on Outputs
  |
  v
Outputs
```

The first system is an expensive approach to quality control. Material, people and equipment resources are used to produce both good and bad product, and at the end of the process more people and equipment are used to filter out the scrap generated. In addition, filtering processes tend not to be 100 per cent efficient, which means customers often receive poor quality product, even after inspection.

The second system has a more proactive approach to quality. Systems are installed to prevent problems within the process. The output is sample-tested because the good quality from the line doesn't necessitate 100 per cent inspection. The findings of the sample test are fed back to the personnel responsible for the process and preventative measures are continuously taken to eliminate any sources of variation. Product quality costs are therefore reduced because only the minimum of materials, labour and equipment processing time have been invested in the production of the goods.

The techniques used to eliminate variation on the production floor can also be used in other areas of the business.

Total Quality Management is the term used to describe a management-led drive for quality throughout the organisation. All activities and functions within the organisation are scrutinised, and improvement opportunities identified.

TQM aims to focus the organisation on both internal and external customer needs. It seeks to introduce quality practices into every area of the business, from research and development to manufacturing, sales, marketing, finance and other departments. Each function and each person within the function identifies their customers, what service they supply to these customers and how the effectiveness and efficiency of that service can be measured. Each function and responsible staff member aims to achieve a "right first time" approach to tasks.

In a manufacturing context, each operation defines what it requires of its supplying operations. This may be as simple as providing units that are clean and not damaged in transit, or it may involve developing tolerances within which the units or services must perform.

Performance targets are identified and a team approach is adopted for achieving these targets. Targets may include, among other things, new product development time to market, delivery reliability improvement, reduction in errors made in software programming, reduction in defects from manufacturing or achieving cost reduction targets.

The tools described in this chapter are presented in the context of the manufacturing environment but are just as applicable to the administrative functions of the business.

MANUFACTURING QUALITY DEFINED

There are two commonly used definitions of quality which are applicable to the manufacturing environment. Each of the definitions contributes to our understanding of the concept of manufacturing quality.

Quality can be defined as *fitness to standard*. Fitness to standard compares the unit produced to the customer's specifications. Various characteristics and functions of the unit are measured against the design specification and if the unit passes, it is considered a quality part.

Another definition of quality is *fitness to cost*. Fitness to cost focuses on ensuring the part is produced within specifications but also that it is produced economically. The cost of the part has to be acceptable to the customer and to the manufacturing company. Fitness to cost means high quality at a low cost. In order to minimise costs, any sources of waste have to be eliminated. The manufacturer has to put in place techniques to avoid scrap generation. Non-value activities such as inspection should, ideally, be eliminated.

Both of these definitions combine to clarify what is meant by manufacturing quality. Fitness to cost ensures the product is affordable by eliminating waste within the production process. Fitness to standard ensures that all parts leaving the production facility satisfy the product specification.

THE COST OF QUALITY, AND THE SUITABILITY OF QUALITY IMPROVEMENT PROGRAMMES

Quality should be factored into all aspects of the manufacturing system. However, many companies are slow to allocate the resources and time required to install quality systems and procedures. This can be overcome by determining the cost of quality as a benchmark to highlight the scope for improvement through the use of quality tools. As the saying goes: "Quality costs, but poor quality costs more!"

The cost of quality can be estimated by imagining what tasks and expenditures would be avoided if raw materials and manufacturing processes were perfect. If raw material arrived in perfect condition consistently, there would be no requirement for incoming inspection. If the manufacturing processes always produced only perfect product, there would be no scrap, no customer complaints, and indeed, no need for a quality department at all. Each of the costs associated with these imperfections can be considered as the cost of quality. They are broken down and categorised under the headings:

- Prevention Costs
- Appraisal Costs
- Failure Costs.

Prevention costs are those incurred in taking steps to avoid problems at their origin. They include the cost of performing pilot production runs for new product introductions, measuring the capabilities of various equipment, installing and maintaining systems such as ISO 9000, to name but a few. Resources allocated to prevention are focused on ensuring that only good units are produced.

Appraisal costs relate to inspection and verification tasks. These include, amongst other things, expenditure on inspecting raw materials, in-process product inspection, quality audits, and assessment of vendors.

Failure costs are generally sub-divided into internal and external failure costs. Internal failure costs relate to the cost of generating defect units which are detected before shipment. These costs include the cost of repairing or scrapping the units, the cost of re-inspection and failure analysis costs.

External failure costs relate to defect units being discovered by the customer in the field. These costs include the loss of goodwill, product liability costs, the cost associated with the repair of the unit and/or the supply of a new unit to the customer. Factories with poor quality strategies suffer from very high failure costs. Failure costs tend to be significantly greater than either prevention or appraisal costs.

Figure 2.2 shows the relationship between the various forms of costs. As a company adopts more preventative tasks, the overall cost of quality is reduced. Prevention is ultimately the cheapest form of quality control.

FIGURE 2.2: THE COST OF QUALITY

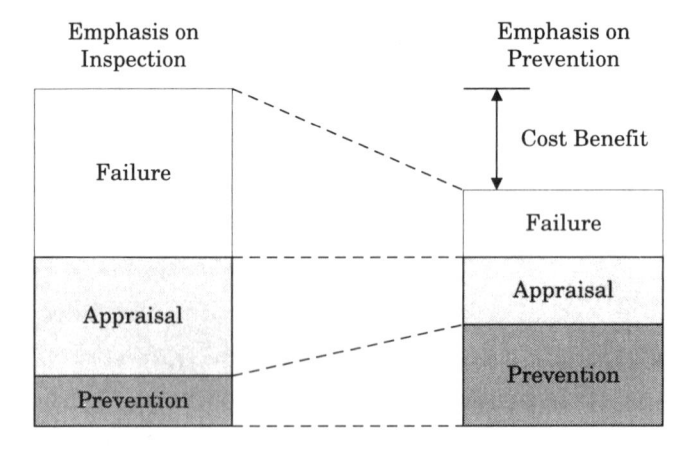

The cost of quality as a percentage of sales or as a percentage of the manufacturing cost can be benchmarked against "best in class" companies to determine the scope for improvement. As Figure 2.2 suggests, investment in prevention can significantly reduce quality costs.

PRELIMINARY WORK FOR TQM INTRODUCTION

A company-wide strategy to implement TQM is a major undertaking in commitment and effort from management. It is important to establish expectations for the improvement programme at the outset. In this way, the success of the programme can be monitored regularly and adapted as needed. The first step is to establish clear goals and to assess if TQM is the most appropriate means of achieving them.

Goals can be established by comparing the performance of the factory to sister plants or competitors. Customers are generally willing to document how they would like performance to improve. A typical range of areas for improvement include:

- Quality of product
- Delivery reliability
- Cost reduction
- Monthly throughput
- Machine maintenance levels
- Customer response time
- New product introduction
- Employee turnover
- Inventory levels
- Inventory accuracy
- Forecast accuracy
- Scrap level generation

Goals will have to be quantified or set against a variety of criteria; for example, product quality to improve by 5 per cent per year or cost to reduce by 10 per cent per year. Top management should provide a clear understanding of the gaps in performance and recognise the potential for the application of quality tools to bridge that gap.

The problem-solving tools are readily applicable to most problems and could be implemented on an isolated basis to achieve specific targets. However, the spirit of TQM is company-wide improvement.

In order to instil improvement across all functions, there must be strong leadership from the top manager. In the scenario of the manufacturing plant, the plant manager who perceives TQM as a strategy for survival must promote the concept throughout the organisation. All employees are mobilised to achieve continuous improvement. The full benefit of TQM is achieved through a company-wide programme which aims to change the philosophy of the organisation.

The organisational structure may have to change to accommodate TQM. TQM will need dedicated management resources to drive the programme. In one factory where I worked, a TQM facilitator was promoted to management level and given a small department from which to organise and promote TQM events. This person worked with management to set in motion the training courses and structures needed to disperse knowledge on improvement tools throughout the organisation.

It is necessary for all employees to participate in the improvement process and therefore it is important that good industrial relations exist between management and staff. Operators will have to be given the opportunities to determine targets, gather quality information and develop improvement plans. Indeed, trust is an essential ingredient in introducing and implementing the concepts of TQM.

Kick-starting the TQM process begins with a training session for managers where the benefits of TQM are explained and the common improvement tools are introduced. These improvement tools can be used to enable improvement teams achieve targets within their work area. They enable control of the variation which causes inconsistencies within the process and they create a framework for continually improving performance. As each team achieves its target, new ones are set and the cycle of improvement continues.

The Improvement Cycle: Plan — Do — Check — Act

Deming taught Japanese companies that there are four cyclical steps to continuous improvement. These four steps have become widely accepted as a means for driving continuous improvement. The cyclical steps are *plan — do — check — act* or PDCA for short.

FIGURE 2.3: THE PDCA CYCLE

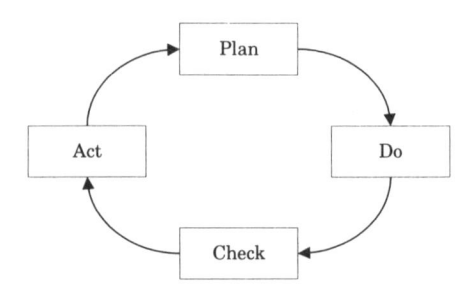

The *plan* step consists of defining and analysing the problem. The planning stage involves quantifying current performance, understanding the history behind this performance level and defining the performance target. Once the problem is clearly defined, solutions can be sought. Take the following example.

A company defines a goal of reducing facility costs by 10 per cent for the factory. The level of facility costs are calculated as £400,000, made up of electricity (£160,000), water (£70,000), gas (£60,000), equipment upgrade (£50,000), miscellaneous (£60,000). Therefore a reduction of £40,000 is the target. Solutions for achieving this target need to be identified.

A team is set up consisting of the users of these overheads, production, maintenance, finance, personnel, and the project leader. The solutions generated include the better use of reduced cost, night rate electricity, recycling of water, recycling of heat generated from some processes, re-negotiation of the contract with the gas supplier. From their early meetings, the team develops an improvement plan from suggestions.

The planning stage concludes when the size of the problem has been quantified and an improvement plan has been agreed.

The *do* step involves implementing the planned tasks. In the above example, some equipment is designated to be used only during night-time hours, the gas company provides a rebate depending on the level of gas used and water is recycled for some equipment.

The *check* step involves evaluating the effects of the action taken with regard to the overall goal. It is the feedback step which measures the effect of work performed. Many companies consider that the completion of the *do* step automatically resolves the problem. The philosophy of continuous improvement uses the check step to verify that the goals have been achieved. The check stage ensures that the correct response and direction in performance are being derived from actions taken. It takes stock of progress made and helps to fine-tune improvement work.

In the above example, a cost reduction of £40,000 was sought and a reduction of 30K was achieved. The company now has a choice of either undertaking extra actions to achieve the final £10,000, or in the spirit of continuous improvement, setting another goal of, say, a £20,000 reduction. Before determining how to achieve the new target, the earlier work must be documented and standardised as the new work method.

The *act* step consolidates work done to date. Procedures are written which ensure changes become part of standard work practices and audits are performed to ensure new instructions are adhered to. In this example, the team writes procedures to ensure that nominated equipment such as ovens are only used at night. As part of the continual improvement process, the act stage returns the team to plan stage to identify new opportunities for cost reduction.

The team in our example returns to the plan stage and decides to negotiate a long-term agreement with equipment suppliers to reduce the equipment upgrade and spares costs. They also define guidelines for equipment selection to ensure that their impact on facility costs is considered as part of the cost justification.

The PDCA approach to improvement is cyclical by nature. The wheel continues to turn as objectives are met and new ones set.

Analysis of Total Quality Management Tools

There are a wide variety of improvement tools available to the improvement team. Each of the tools helps the team to identify sources of variation in the process and to control it. The elimination of variation in the process introduces stability and consistent output. By progressively reducing variation and improving the capability of the process, the level of defects continuously declines and productivity increases. The following tools are described under two headings: problem identification and analysis tools; and statistical techniques.

Problem identification and analysis tools include:

- Flow Charting
- Checksheets
- Pareto Diagrams
- Brainstorming
- Cause-and-Effect Diagrams
- Failure Modes and Effects Analysis

- Checklists.

Statistical techniques include:

- Histogram
- Statistical Process Control (SPC)
- Process Capability studies.

These tools can help the factory to improve its daily performance. They help practitioners to identify critical problems, analyse the source of the problem and develop improvement plans. The statistical techniques continuously eliminate variation in output and enable processes to approach the ideal output characteristics. As variation is eliminated, waste is reduced, costs are reduced and competitiveness improved.

Problem Identification and Analysis Tools

Flow Charting

Flow charting is a pictorial summary showing all the steps in a system. It presents all the tasks that combine to form the system or process in a block diagram format. The diagram displays the interdependencies between tasks. Flow charting begins by breaking the process down into its component parts and representing them using the symbols shown in Figure 2.4.

FIGURE 2.4: FLOW CHARTING SYMBOLS

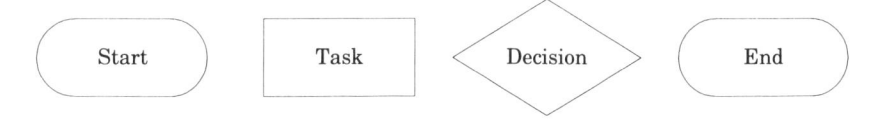

| Start | Task | Decision | End |

Flow charting helps the improvement team by summarising the process. It may reveal discrepancies in the process that were not previously realised. For example, the following flow chart describes an operator performing a standard work cycle on a press.

FIGURE 2.5: FLOW CHART OF A SERIES OF TASKS AT A WORKSTATION

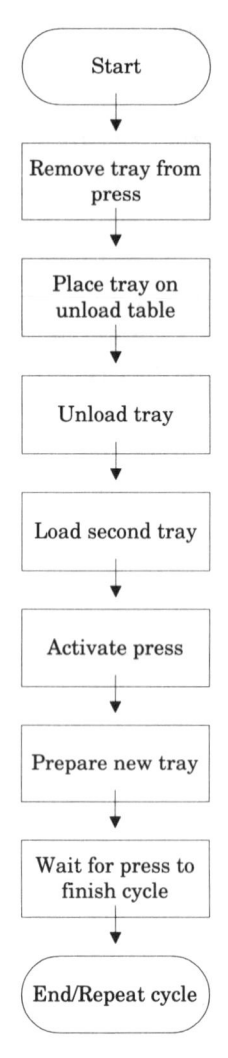

```
        Start
          │
          ▼
  Remove tray from
       press
          │
          ▼
   Place tray on
   unload table
          │
          ▼
    Unload tray
          │
          ▼
  Load second tray
          │
          ▼
   Activate press
          │
          ▼
  Prepare new tray
          │
          ▼
  Wait for press to
    finish cycle
          │
          ▼
   End/Repeat cycle
```

This flow charting of the work cycle reveals that by moving the "unload tray" step to after the "activate press" step, productivity will improve for the operation. This is because the press, not the operator, is the bottleneck in the cycle; therefore moving this task improves its utilisation. Figure 2.6 shows a flow chart for an electronics assembly line.

FIGURE 2.6: FLOW CHART OF AN ELECTRONICS ASSEMBLY LINE

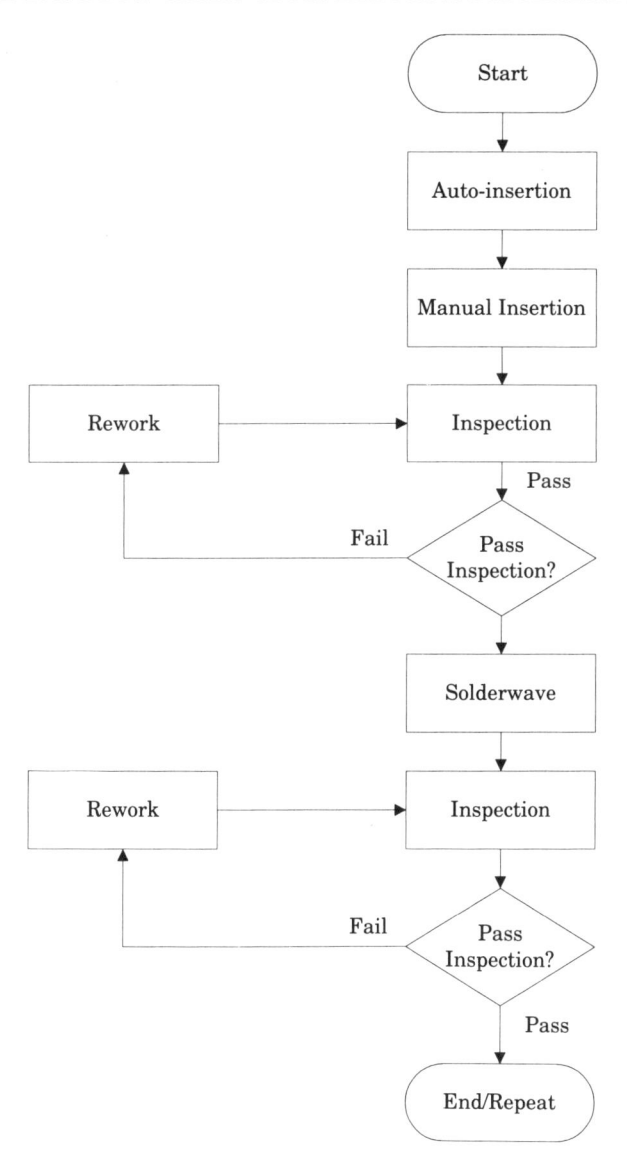

```
                    Start
                      ↓
              Auto-insertion
                      ↓
             Manual Insertion
                      ↓
   Rework  →     Inspection
     ↑                ↓ Pass
     |         Pass Inspection?
     ←── Fail ──┘
                      ↓
                 Solderwave
                      ↓
   Rework  →     Inspection
     ↑                ↓
     |         Pass Inspection?
     ←── Fail ──┘
                      ↓ Pass
                 End/Repeat
```

In summary, flow charting provides an overview of what is happening in a process. The chart enables the team to gain an increased perspective on the interdependencies between the various tasks and should assist decision-making.

Checksheets

Checksheets are forms that are used for data collection. They are simple to use and provide information regarding the frequency in which various problems occur. They are a useful first step in beginning a prevention programme on a process because they provide factual data regarding the extent of various problems.

A checksheet may be designed as in Figure 2.7, where a tally mark indicates the occurrence of the problem. Figure 2.8 illustrates an alternative design for a check sheet showing the frequency of the problem and also the location on the product where it occurred. In this case, the operator is given a picture of the part and marks the position of the problem each time it occurs. The checksheets indicate both the frequency and location in which the problem occurs.

FIGURE 2.7: CHECKSHEET FOR A PRINTED CIRCUIT BOARD (PCB) AFTER A SOLDER WAVE OPERATION

Fault	Mon	Tues	Wed	Thurs	Fri	Total
Excess Solder	ЖН	ЖН ЖН	III	IIII	ЖН III	30
Unsoldered Devices	ЖН ЖН ЖН ЖН IIII III		ЖН ЖН ЖН ЖН ЖН	ЖН ЖН ЖН ЖН	ЖН ЖН	82
Broken Tracks	III	ЖН II	IIII	III	IIII	21
Raised Devices	IIII	IIi	III	III	II	15

Figure 2.8 shows how scratches occur in a variety of places on the housing but are concentrated in the top right corner. By investigating the cause of the scratches in this area, the problem is much reduced. The information gathered by the checksheet can be organised into a Pareto diagram for ease of interpretation.

Total Quality Management Tools 53

FIGURE 2.8: CHECK SHEET FOR MONITORING SCRATCHES ON A HOUSING

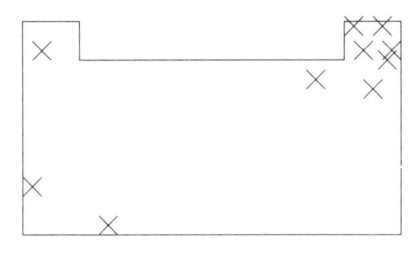

Pareto Diagram

When an operation contains many problems which all need attention at the same time, the improvement team must concentrate their efforts on the most severe ones. A Pareto diagram enables the problem-solving team to prioritise problems according to their severity. The Pareto diagrams concentrate the team's attention on the "vital few" rather than the "trivial many".

Essentially, the Pareto diagram is a bar chart where the bars are arranged in decreasing order from left to right. The raw data for diagrams is collected on checksheets and then converted into Pareto format. Figure 2.9 shows a Pareto diagram constructed from the checksheet in Figure 2.7.

FIGURE 2.9: PARETO DIAGRAM OF THE FAULTS ON THE PCB

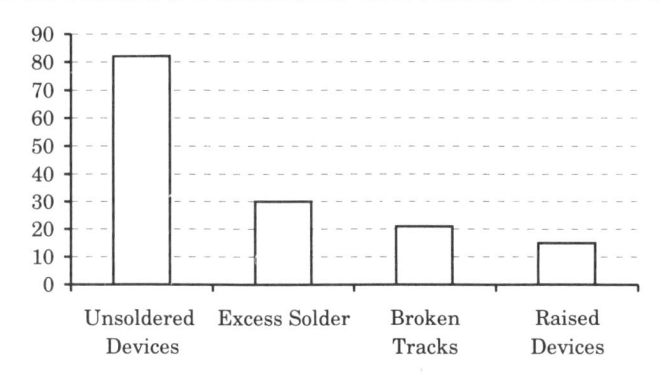

The diagram shows that 55 per cent of the problems will be resolved if the team successfully resolves the unsoldered devices problem. Brainstorming is an effective means of generating ideas on why problems happen and how they may be avoided.

Brainstorming

Brainstorming was conceived by Alex Osborn in 1939 to help his advertising agency develop creative promotional ideas. Osborn's staff would "storm" the problem with their "brains". The same techniques help manufacturing personnel develop innovative problem solving ideas.

Brainstorming consists of inviting approximately 5–8 people who have expertise pertaining to a manufacturing problem, to contribute ideas about possible causes of the problem. Osborn defined guidelines and rules to help generate the maximum number of ideas.

- Organise a group of people who are familiar with the problem and who are capable of generating many ideas on its possible causes.
- Ensure people have sufficient notice of the brainstorming session and its aims so that they have an opportunity before the meeting to consider the background facts and potential causes.
- Repeatedly go around the table requesting one idea per person. This method of requesting ideas from each person encourages the timid members to contribute and helps control over-enthusiastic people.
- The aim of the session is to generate the maximum number of ideas; therefore no criticism of ideas is allowed. Analysis of ideas is left until much later;
- Record the ideas on a flipchart; continue the process until all ideas have been generated.

The team, in our example, consists of the area supervisor, two operators, the machine technician, and the process engineer. Through a brainstorming exercise, the following list of possible causes of the unsoldered devices is generated:

- The flux cleaning agent may not be strong enough
- Solder contamination
- Product design incompatible with solder wave

- Too many product changeovers affect machine settings
- Dirt on machine's preheaters
- Difficult to load machine conveyor correctly
- Insufficient time allocated for cleaning and tidying the machine
- Angle of conveyor
- Quantity of solder in the solderbath
- Handling components causes contamination
- Poor conveyor maintenance
- Air knife blowing solder away from devices
- Poorly inserted devices
- Variation in environment temperature
- Variation in factory extract system.

Brainstorming can generate a broad list of causes that need to be analysed and controlled. Sometimes it can be difficult or daunting to know where to begin. A cause-and-effect diagram is an easy way of categorising. This categorisation of causes makes it easier to analyse them.

Cause-and-Effect Diagrams

Professor Kaoru Ishikawa developed cause-and-effect diagrams as a tool for evaluating the causes of problems. Ishikawa found that factory employees were often overwhelmed by the number of causes related to problems. The cause-and-effect diagram lets them categorise the causes under five headings: People, Method, Material, Machine and Environment.

A cause-and-effect diagram consists of the main trunk leading to the problem statement or the "effect" and five branches representing each category of cause. The list of causes for unsoldered devices can be distributed under these headings. The team analyses each cause and decides if it is primarily a machine, materials, method, people or environment issue. Figure 2.11 shows how the list is distributed.

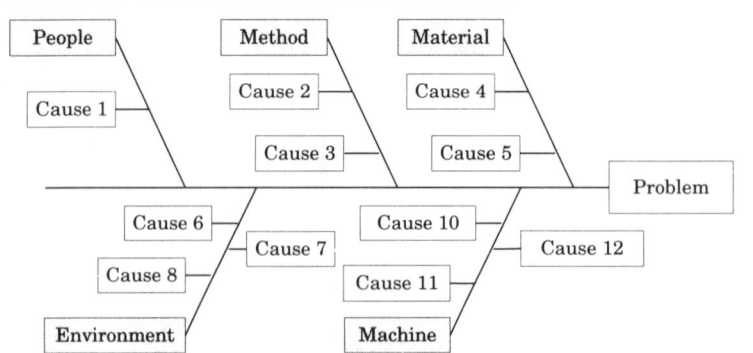

FIGURE 2.10: CAUSE-AND-EFFECT DIAGRAM

[Fishbone diagram with branches labeled People (Cause 1), Method (Cause 2, Cause 3), Material (Cause 4, Cause 5), Environment (Cause 6, Cause 7, Cause 8), Machine (Cause 10, Cause 11), and Cause 12, all leading to Problem.]

FIGURE 2.11: CAUSE-AND-EFFECT DIAGRAM FOR UNSOLDERED DEVICES EXAMPLE

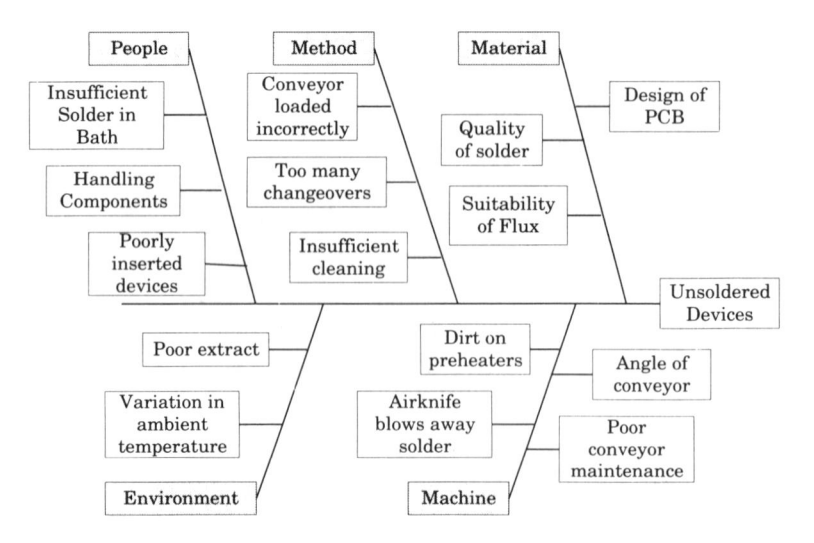

[Fishbone diagram with branches:
- *People: Insufficient Solder in Bath, Handling Components, Poorly inserted devices*
- *Method: Conveyor loaded incorrectly, Too many changeovers, Insufficient cleaning*
- *Material: Quality of solder, Design of PCB, Suitability of Flux*
- *Environment: Poor extract, Variation in ambient temperature*
- *Machine: Dirt on preheaters, Airknife blows away solder, Angle of conveyor, Poor conveyor maintenance*
- *leading to Unsoldered Devices]*

Categorisation in this way enables the team to obtain a better understanding of the causes. It prepares the team for the next step — performing a *failure modes and effects analysis*.

Failure Modes and Effects Analysis (FMEA)

Failure modes and effects analysis is a team-based problem-solving technique. The objective of the exercise is to help the improvement team to develop problem prevention techniques.

FMEA is a natural progression from the earlier stages of problem-solving, namely brainstorming and cause-and-effect analysis. It is used to develop preventative methods that help to minimise the occurrence of process problems. An FMEA systematically summarises all information relating to a problem and prioritises areas that need solutions — for example, an engineering or R & D solution is needed that may be beyond the capabilities of the team.

A failure modes and effects analysis numerically quantifies the problem under the following headings:

- How significant is the problem?
- How effectively are its causes controlled?
- What improvement action is being undertaken?
- What will be the beneficial effect of the improvement?
- What action should be taken for out-of-control conditions?

These are the questions that spring naturally to the mind of any manufacturing engineer who has to resolve manufacturing problems. When the answers are identified and written down in an understandable format, the information is readily available for training operators, technicians, engineers, etc. Figure 2.12 illustrates an example of an FMEA with "unsoldered devices" as the failure mode.

The headings for each column are described as follows:

1. *Potential Failure Mode*: Define the manner in which the process may potentially fail to meet its requirements. In the example relating to the solder wave machine, unsoldered devices are considered as the failure mode and subsequent questions are answered with respect to this problem.

2. *Potential Effects*: Describe the possible effects the failure mode could have on the customer. There are often many potential effects caused by the same failure mode. In the case of unsoldered devices, the printed circuit board will fail electronic test and have to be repaired. If the product succeeds in passing electronic test, it will most likely fail in the field during its early-life stage.

3. *Severity*: describes how significant the potential effect of the failure is on the customer. This is quantified on a scale of 1

to 10 where the values relate to how seriously the individual company perceives each potential effect — 1 indicates "no effect" and 10 indicates "may endanger machine operator or end customer". For example, in Figure 2.12, the company perceives customer returns as a more serious problem than either test rejects or reworking of devices.

4. *Causes of Failure*: lists the causes associated with each failure mode. The causes that have been identified during brainstorming and the cause-and-effect analysis are listed.

5. *Occurrence*: the frequency with which each specific cause of failure is likely to occur. This is numerically gauged where a typical scale has 1 for remote possibility of occurrence (e.g. 1 in 1,500,000), and 10 for very high probability of failure due to the cause (e.g. 1 in 2). This scale can be interpolated to provide intermediate values.

6. *Current Controls*: describes methods used to contain the causes of failure. This may include the use of foolproofing or a checklist that verifies that parameters are at their correct setting before commencing production.

7. *Detection*: numerically defines how capable the controls are at containing the problem. This question seeks to quantify the likelihood of the control preventing the fault getting to the customer. On a scale from 1 to 10, 1 indicates that the controls will almost certainly detect the failure mode, 10 indicates that the current controls are unable to detect the failure mode and thereby prevent it getting to the customer.

8. *Risk Priority Number (RPN)*: is a product of Severity (S), Occurrence (O) and Detection (D), i.e. RPN = S x O x D. The risk priority number summarises the potential severity of the process's problems, the frequency of their occurrence and the capability to contain the problem. In effect it describes the company's exposure to problems in each process. The company should aim to continuously minimise the RPN number for all processes, initially setting a target for every process to be under 100 points, and eventually under 10 points.

FIGURE 2.12: EXAMPLE OF FAILURE MODES AND EFFECTS ANALYSIS FORM

Process Description	Potential Failure Modes (1)	Potential Effects (2)	Severity (S) (3)	Cause of Failure (4)	Freq. of Occurrence (O) (5)	Current Controls (6)	Detection (D) (7)	RPN (SxOxD) (8)	Recommended Improvements (9)
Solderwave	Unsoldered Devices	Causes test rejects	5	Flux cleaning agent	5	Replace flux density	3	105	Check flux density every 4 hours
		Rework devices	5	Solder contamination	2	Replace solder every 2 months	3	42	Analyse solder content every month
		Customer Returns	5	Product design	1	Engineering inspect product drawings	5	35	Test prototypes on machines
				Frequent Product Changeover	2	Operator training	4	56	Define a limit of number of changeovers per day
				Insufficient cleaning	5	Cleaning time allocated	5	175	Increase time of cleaning
				Incorrect loading of conveyor	2	Operator training	2	28	Retrain operators
				Angle of conveyor	2	Operator training	3	42	Mark correct position per product on machine
				Etc.	Etc.	Etc.	Etc.	Etc.	Etc.

9. *Recommended Action*: this column describes the action to be taken to reduce the RPN number. Corrective action is directed to factors that have high RPN. The exercise of performing an FMEA helps to concentrate the team's efforts on the major causes generating the problem.

The FMEA is a concise documented summary of the problem as analysed by the process engineer and production staff. It draws attention to the controls used to minimise the various causes of process problems. Checklists and statistical process control are very versatile tools for controlling the causes of problems.

Preventing Potential Problems with Checklists

Before a pilot heads down the runway to take his passengers to their holiday destination, a defined check is carried out on all instrumentation to verify it is in correct working order. No doubt all passengers feel more confident about the safety of their journey knowing that a preliminary check has been performed before the flight begins. The same philosophy holds true for the factory machine operator; by beginning the shift with a predefined check on the process set-up, there is a much improved probability of having a trouble-free day.

Checklists are designed to specify which aspects of the process should be inspected before beginning production. The areas which need inspection and the inspection criteria are derived from the FMEA. Figure 2.13 shows a checklist for a solderwave machine.

Introduction to Statistical Techniques

Histogram

Deming's approach to quality control was to reduce the variation inherent within the process and as a result reduce the variation in product output. There are multiple causes of variation, as indicated by the cause-and-effect diagram earlier. The variation in the solderwave output's quality depends on such factors as the density of the flux, the speed of the conveyor, the temperature of the preheaters or the quality of the components

FIGURE 2.13: CHECKLIST FOR PREVENTING UNSOLDERED DEVICES

Process Checklist: Solderwave   Operator:

Parameter	Criteria	Frequency	Mon	Tues	Wed	Thurs	Fri	Sat
					Week Date			
Clean conveyor fingers	No flux residue	End of Shift	Yes	Yes	Yes			
Density of the flux	0.9–0.95	Every 4 hours	0.92 0.99	0.90 0.94	0.95 0.91			
Conveyor width	Check product spec	At changeover	10	10	10			
Conveyor speed	Check product spec	At changeover	2.1	2	2.1			
Preheaters clean	No flux residue	End of shift	Yes	Yes	Yes			
Temperature of preheaters	120°C–130°C	Start of shift	125°	122°	123°			
Solder bath temperature	220°C–225°C	Start of shift	221°	222°	220°			
Angle of conveyor	Check product spec	At changeover	7 deg.	7 deg.	7 deg.			
Etc.								

being soldered. As each of these factors varies, the quality of the output from the process also varies. There are several statistical techniques available for analysing variation and improving our understanding of the production process. The histogram is a commonly used technique that provides a graphical interpretation of variation within either the production process or within the product itself.

A histogram converts measurement data into a bar chart, which represents the frequency distribution of the data. The graphical representation is much easier to comprehend and generally provides an insight into the process performance. Constructing a histogram begins with a table of data regarding the characteristic under investigation. The data is gathered randomly over a period of time. Table 2.2 represents variation in bracket dimensions after a forming process. The customer specification for the part is 9.85mm to 10.15mm. The objective is to convert this raw data into a frequency bar chart that readily reveals information regarding variation within the process, the central tendency of the process and the shape of the distribution curve.

TABLE 2.2: BRACKET DIMENSIONS

Specification: 9.85–10.15 mm			
10.04	10.07	9.97	9.98
10.00	10.00	10.02	10.05
10.00	10.11	10.12	9.96
9.90	10.00	9.95	10.03
10.01	9.95	10.01	10.05
10.00	9.90		

The range, R, for the data is calculated by subtracting the smallest value from the largest value within the data. In the above example, the range is calculated as:

$$R = 10.12 - 9.9 = 0.22 mm$$

This range is divided into intervals, and the frequency of occurrence of the characteristic within each interval is determined. A rule of thumb for determining the number of intervals is to equate it to the square root of the sample size. Table 2.3 shows a frequency table based on the values provided in Table 2.2. Figure 2.14 shows the histogram derived from the frequency table.

TABLE 2.3: FREQUENCY TABLE FOR BRACKET DIMENSIONS

Interval	Frequency
9.9 – 9.944	2
9.945 – 9.988	5
9.988 – 10.032	9
10.032 – 10.076	4
10.076 – 10.120	2

FIGURE 2.14: HISTOGRAM OF BRACKET DIMENSIONS

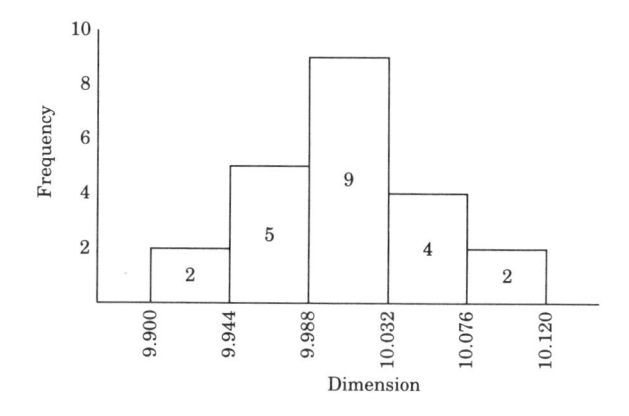

The histogram shows the central tendency or location of the distribution. It has an average or mean of 10mm. The spread of the distribution is from 9.8 to 10.2mm and is approximately symmetrical about the mean. The distribution is bell-shaped. This shape of distribution is very common; many natural and technical processes that occur randomly can be described by the *normal distribution*. The histogram has converted the tabular data into a pictorial representation that is easy to understand, and describes the performance of the process in relation to the product specification.

The Normal Distribution

The normal distribution describes the frequency of occurrence of a random variable. The distribution can be mathematically defined by its average or mean (μ) and its standard deviation (σ). The standard deviation is a measure of the variability within the process; the greater the variability, the greater the spread of measurements and the larger the standard deviation. The following probability density function mathematically describes the normal distribution of a random variable x:

$$f(x|\mu,\sigma^2) = \frac{1}{\sigma\sqrt{2\pi}} \cdot e^{\frac{-(x-\mu)^2}{2\sigma^2}}$$

The probability that a product, taken randomly from a normally distributed population, will have its variable characteristic X within a distance d from the mean can be calculated by integrating the above formula.

$$F(x|\mu,\sigma^2) = P(X \leq x) = \int_{-\infty}^{x} f(x|\mu,\sigma^2)dx$$

Table 2.3 shows the percentages calculated for distances of +/-1σ, +/-2σ, +/-3σ. Figure 2.15 shows how these percentages can be interpreted on the normal distribution curve.

TABLE 2.4: PROBABILITY THAT A NORMALLY DISTRIBUTED RANDOM VARIABLE X IS LOCATED WITHIN THE INTERVAL μ+/-d

Distance d	Probability p
+/−1σ	68.26%
+/−2σ	95.44%
+/−3σ	99.74%

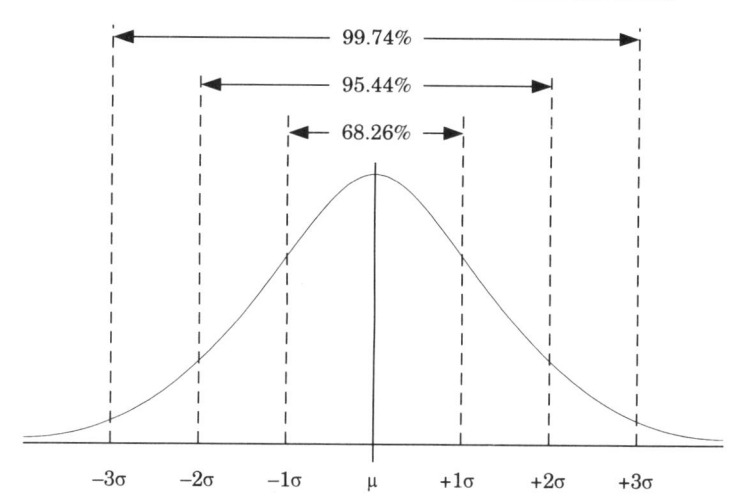

FIGURE 2.15: RELATIONSHIP BETWEEN THE STANDARD DEVIATION AND THE VARIABILITY WITHIN THE PROCESS

This mathematical interpretation of the normal distribution enables us to further analyse the data provided in Table 2.2 and the histogram in Figure 2.14. The mean of this distribution is calculated as 10mm, the standard deviation as 0.05mm. From these parameters, it can be concluded that 99.74 per cent of brackets have dimensions between:

$$(\mu+3\sigma) - (\mu-3\sigma) = 10.16 - 9.84$$

Therefore 26 units in every 10,000 or 2,600 units in every million produced (100.00 − 99.74 = 0.26) will be outside the customer's specification. Thus, the mathematics of the normal distribution enable us to predict the level of defects being sent to the customer from samples of data taken randomly from production. While 2,600 defects per million may be acceptable to some customers, many others may insist on quality defects as low as 60 defects per million. Statistics plays an important role in attempting to achieve these quality levels.

The histogram provides a graphical image of the variability of process output. In general, it is used to interpret historical data and provide information for an improvement team. It enables them to experiment with changes and evaluate the results. While a histogram provides a graphical analysis of historical data, a run chart can be used as a real-time monitor of process performance.

Run Chart

Quality data can be represented on a continuous basis by means of a run chart. Run charts are easy to construct, use and interpret. They provide a graphical representation of process performance over time. They can also be used to monitor a range of other factors such as ambient humidity, productivity levels, maintenance levels or product quality characteristics such as bracket dimension, as in Figure 2.16. A random sample of brackets is taken at regular time intervals, as in the accompanying table below, and the average bracket dimension is plotted on a chart. The chart describes the performance level for the bracket forming process. Continuous deterioration or improvement of the process will be recognised from the chart and further analysis can be undertaken.

A second chart can be used to monitor the range of measurements for each sample. This chart plots the difference between the minimum and maximum measurement within each sample and monitors the sample's variability. It ensures that high variation above and below the average within the one sample is highlighted. Figure 2.17 shows the range graph derived from the example in Figure 2.16.

The run chart provides a valuable source of information for failure analysis staff who have to investigate the causes of product failure. It describes the performance of the process during various time periods. Damaged parts can be traced back to the time when they were produced and the run chart can be analysed for clues regarding the source of failure. In general, users of run charts encourage extra information to be gathered on the run chart over time — for example, the raw material batch being used during the time period, the operator working on the process, any maintenance work undertaken. All this information is then available to failure analysis staff for their investigation.

Figure 2.16: Run Chart for Monitoring the Bracket Dimension

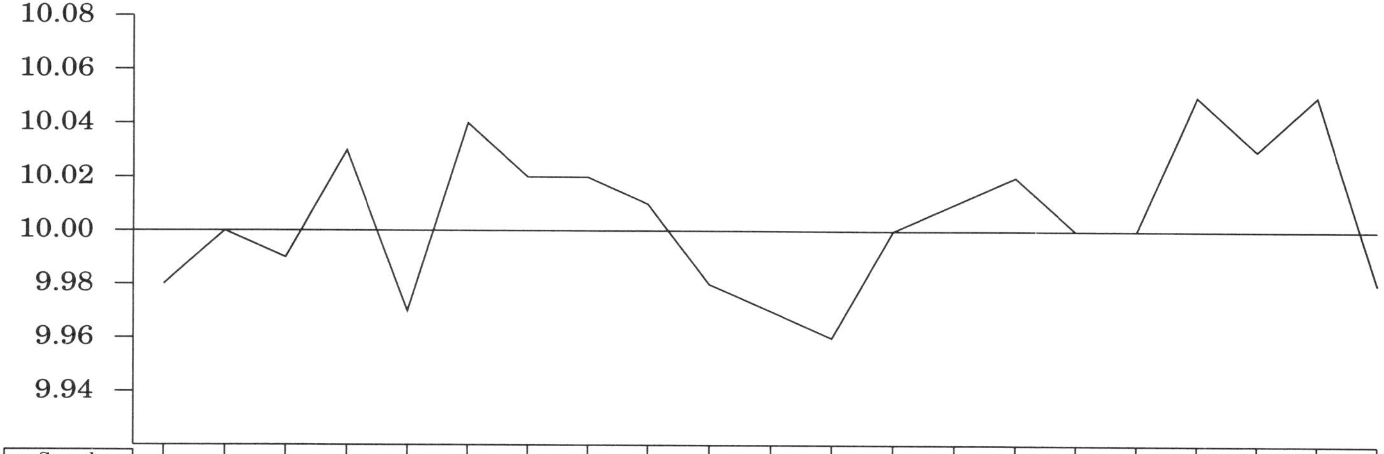

Sample Number	1	2	3	4	5	6	7	8	9	10	11	12	13	14	15	16	17	18	19	20	21
1	10.00	9.99	9.99	10.03	9.96	10.02	10.01	10.04	10.02	10.00	9.98	9.95	10.01	10.02	10.01	10.00	9.98	10.04	10.04	10.03	9.99
2	10.02	10.01	9.98	10.03	9.95	10.06	10.00	10.00	10.01	9.99	9.97	9.96	10.00	10.02	10.03	9.98	9.99	10.05	10.03	10.04	10.00
3	9.96	10.05	9.97	10.01	9.99	10.00	10.03	10.00	10.01	9.96	9.96	9.95	9.99	10.00	10.04	10.02	10.02	10.06	10.01	10.06	9.96
4	9.94	9.95	10.02	10.05	9.98	10.08	10.04	10.02	10.00	9.97	9.97	9.98	10.00	9.99	10.00	10.00	10.01	10.05	10.04	10.03	9.97
X̄	9.98	10.00	9.99	10.03	9.97	10.04	10.02	10.02	10.01	9.98	9.97	9.96	10.00	10.01	10.02	10.00	10.00	10.05	10.03	10.04	9.98
R	0.04	0.10	0.04	0.04	0.04	0.08	0.04	0.04	0.02	0.04	0.02	0.03	0.02	0.03	0.03	0.04	0.04	0.02	0.03	0.03	0.04

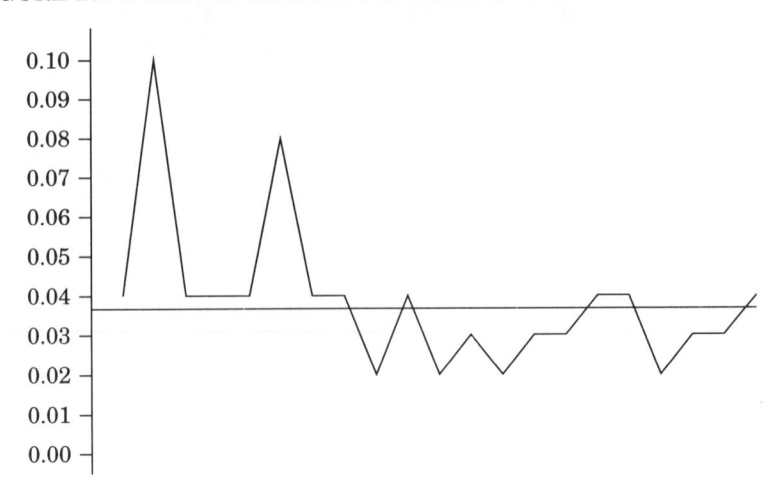

FIGURE 2.17: RANGE GRAPH FOR BRACKET DIMENSIONS

A drawback of run charts is that they do not show when to take corrective action if the product quality begins to deteriorate. Taking action after the quality goes out of specification is too late. Scrap product has been generated and may be sent to the customer. Some practitioners use their experience to define action points or to draw action lines; if the quality characteristic deteriorates beyond these points, corrective action must be taken. Drawing action lines in this manner is very subjective and open to confusion and error. When the run chart represents a distribution similar to the normal distribution described earlier, the appropriate action lines can be calculated mathematically.

Action lines or control lines are calculated at +/−3 standard deviations from the process average. These lines represent the limits within which approximately 99.74 per cent of data points should be located. For a consistent process, data points exceeding these control lines should be very rare. If a data point exceeds these limits, then the cause of the occurrence should be investigated.

Statistical Process Control

The 99.74 per cent range of the normal distribution is defined as the *lower control limit* (LCL) and the *upper control limit*

(UCL). These limits are mathematically defined action lines. Simply stated, any variation in performance, either below the LCL or above the UCL, is indicative of high levels of variation within the process and should be investigated to determine the cause. When these limits are drawn on a run chart, such as the one in Figure 2.16, we have the basis of a statistical process control chart.

The data for the SPC chart is gathered in a similar manner as that for the run chart. Twenty or more samples are taken at random from the process at regular intervals of time. Generally, sample sizes vary from four to 10 units per sample. These units are used to calculate the upper and lower control limits by using the formulae in Appendix A.

FIGURE 2.18: STATISTICAL PROCESS CONTROL CHARTS FOR BRACKET DIMENSION

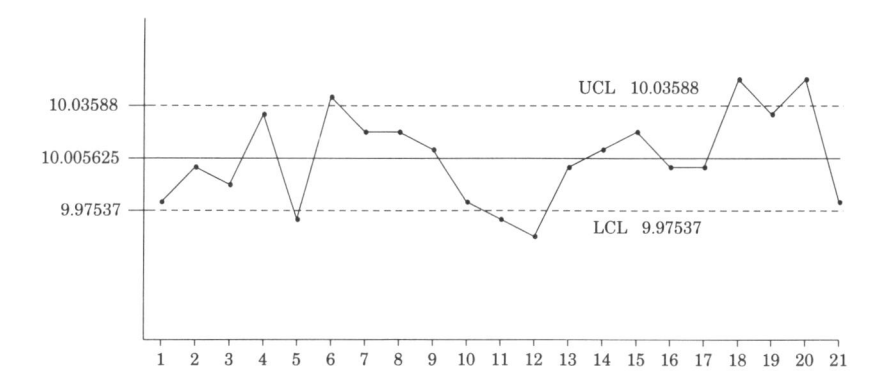

SPC charts should be used to control product or process characteristics which show a high degree of variability and are considered by customers and/or internal staff to be important. The charts can be used to reduce the variability and to control the characteristic.

On a consistently stable process, one would expect data points to be located close to the average, and evenly distributed above and below the average line. Therefore, the following trends are generally considered to be statistically significant and warrant further investigation:

- Any data point outside the upper or lower control limit. In Figure 2.18, there were three points below the LCL and three points above the UCL. Each of these occurrences should be thoroughly investigated at the time of occurrence and action taken to improve the process.
- Seven or more consecutive data points, all of which are either above or below the average line.
- Seven or more consecutive data points continuously increasing or decreasing. Figure 2.18 had seven decreasing points from data point 6 to data point 12. This series of decreasing points is indicative of a trend and warrants further investigation of the causes. The following diagrams provide examples of trends on SPC charts.

FIGURE 2.19: TRENDS ON SPC CHARTS WARRANTING INVESTIGATION

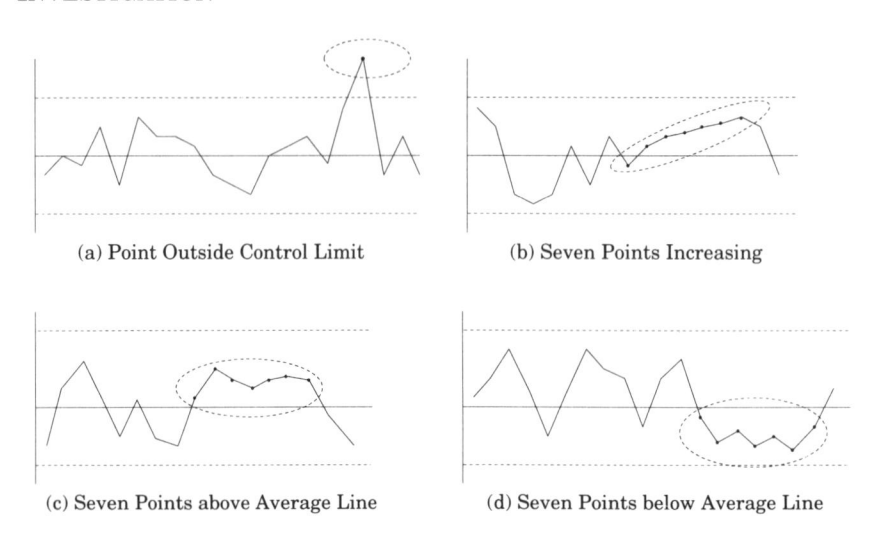

(a) Point Outside Control Limit

(b) Seven Points Increasing

(c) Seven Points above Average Line

(d) Seven Points below Average Line

The sources of variability trends can be divided into two categories: common cause and special causes. Common causes of variation are the multiple sources of variation which are always present within the process. They tend to be an inherent part of the process design and can be very difficult to minimise. They include poor equipment design capability, poor initial product design, and choice of raw materials, among others.

Common sources of process variation are constantly present and therefore tend to be predictable. They are generally responsible for the natural variation of the process around its mean.

Special sources of variation are irregular in their occurrence and therefore unpredictable. They have a strong destabilising influence on the quality from the process and are the main reasons for out-of-control performance. Special sources of variation include inexperienced operators, unplanned events such as power failures, and incorrect equipment repair. For example, a technician may not replace a forming tool correctly after a preventative maintenance session. The machine produces out-of-specification parts until it is highlighted by the SPC chart. Once the source of variation is identified, it is generally easy to correct the problem by training or establishing new procedures.

By investigating each trend on the SPC chart, the team gradually identifies and eliminates each source of variation. Special causes are generally resolved quickly while common cause variation takes time and engineering exertion to overcome. As these causes of variability are progressively eliminated, the process continuously improves.

Capability Studies, Cp and Cpk

Statistical process control charts introduce stability into the process by focusing improvement attention on the sources of variation. As the sources of variation are controlled or eliminated, the process becomes more robust and reliable. Statistical process control charts continuously improve the capability of the process.

Capability indices have been developed to measure a process's capability to produce parts. The Cp index enables the user to determine the capability of a process with respect to their customers' specifications. The Cp capability index of the process can be relatively easily quantified. It is calculated as the ratio of the range between the upper and lower product specifications to six times the standard deviation for the process, i.e.,

$$Cp = \frac{\text{Upper Specification Limit - Lower Specification Limit}}{6 \times \text{Standard Deviation}(\sigma)}$$

$$= \frac{\text{Specification Tolerance}}{6\sigma}$$

The standard deviation measures the spread or variation within the distribution; the process capability index therefore defines how well variation within the process is placed within the specification limits. (The formula for standard deviation is provided in Appendix A.)

FIGURE 2.20: REPRESENTATION OF $Cp < 1$, $Cp = 1$ AND $Cp > 1$

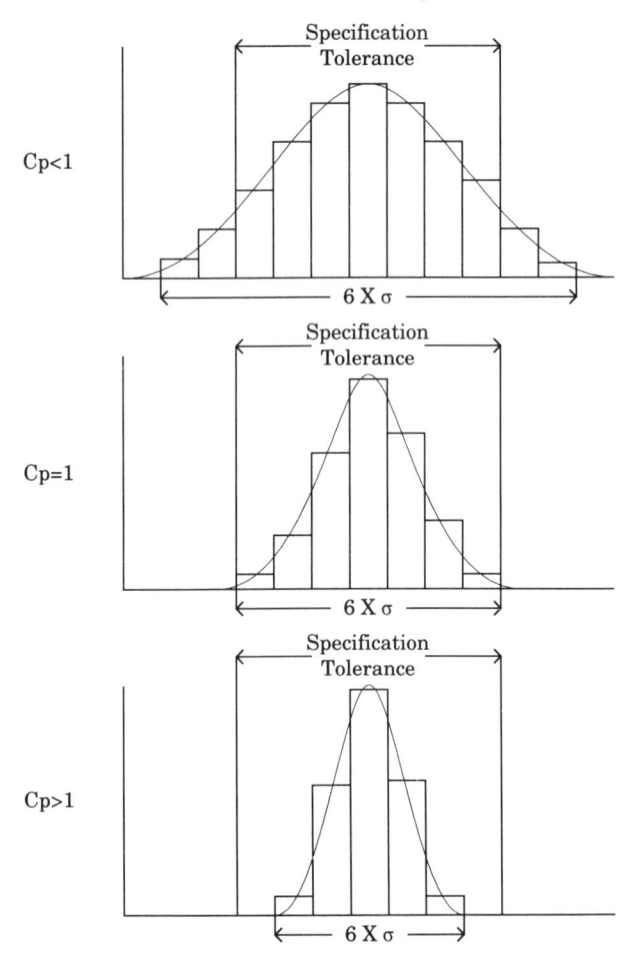

Total Quality Management Tools

The higher the Cp level, the higher the capability of the process to produce units within the customers' specifications. The customer may decide at what level of Cp they consider a process to be capable — for example, they may specify that all processes must have Cp levels above 1 or 1.5, etc.

Knowing the Cp of every factory process gives a good indication of their ability to meet the customer specifications. In Figure 2.18 above, the process capability or Cp is calculated as 1.95, therefore the process appears very capable. The variation in output is approximately half the variation allowed in the product specification.

However, the Cp index has a drawback, it does not differentiate between the capability of two machines as demonstrated in the following example.

FIGURE 2.21: THE CP INDEX FOR TWO DIFFERENT MACHINES

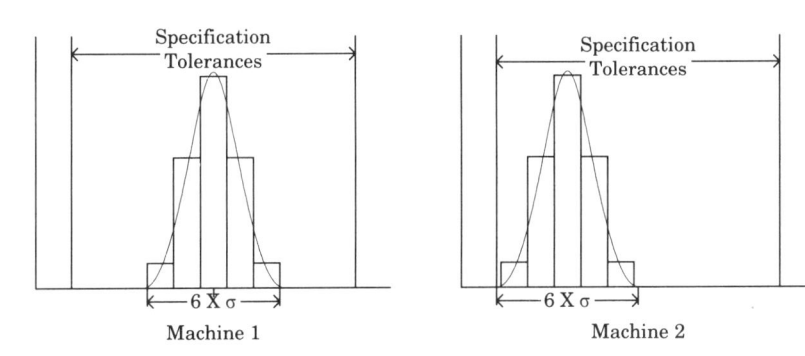

The Cp of both these machines is calculated as 2. However, units from machine 1 are closer to the optimum of 10.0. This characteristic can be recognised in an index called the Cpk index. The Cpk is an enhanced version of the Cp which takes into account the distribution in sizes with respect to the optimum value. Both machines have a Cp of 2 but the average size of product from machine 1 is equal to the specification's average, while machine 2 has an average size closer to the lower product specification. Machine 1 is performing better than machine 2, but this is not reflected in the Cp value. The Cpk index takes this variation into account and is calculated as:

Cpk = the smallest of either:

$$\frac{\text{Upper Specification - Process Average}}{3 \times \text{Standard Deviation}(\sigma)}$$

or

$$\frac{\text{Lower Specification - Process Average}}{3 \times \text{Standard Deviation}(\sigma)}$$

Accordingly, since machine 1 produces product with an average equal to the specification average, its Cpk = Cp = 2. However, machine 2 produces product with an average dimension closer to the specification lower limit and therefore its Cpk is calculated as 1. The Cpk index numerically shows that machine 1 is more capable than machine 2. It directs engineering resources towards those machines requiring the most attention.

Gauge Capability Analysis

An important application of statistical techniques is in evaluating the suitability of gauges for the tasks they perform. In any measurement data, there are two sources of variability present: variability within the parts being measured and variability caused by the measurement system. A gauge with a high level of variability is inappropriate for precision measurements. Statistics enable us to establish the level of variability inherent in the measurement system and thereby establish its suitability for the task. The variability can be expressed mathematically as

$$\sigma^2(\text{total}) = \sigma^2(\text{product}) + \sigma^2(\text{gauge})$$

where $\sigma^2(\text{total})$ is the total variability observed, $\sigma^2(\text{product})$ is the variability due to the product and $\sigma^2(\text{gauge})$ is the variability due to the gauge. Statistical process control charts are used to assess the variability of the gauge.

In the following example, a vernier calipers is used to measure twenty components. The object of the exercise is to establish whether or not the vernier calipers is capable of measuring

Total Quality Management Tools

the component to the required accuracy. Each component is measured twice by the inspection operator who is responsible for performing the task on a regular basis. Table 2.5 shows the data gathered from the measurements and Figure 2.22 shows the $\overline{X}$ and $\overline{R}$ charts.

In this example, the $\overline{X}$ charts show the gauge's ability to distinguish between units. The $\overline{R}$ chart describes the magnitude of measurement error, since the values represent the differences between measurements taken on the same units using the same instrument. The $\overline{R}$ chart is in control; this

TABLE 2.5: SAMPLE GAUGE MEASUREMENTS

Sample	1st measurement	2nd measurement	$\overline{X}$	$\overline{R}$
1	2.38	2.39	2.385	0.01
2	2.37	2.36	2.365	0.01
3	2.38	2.37	2.375	0.01
4	2.37	2.38	2.375	0.01
5	2.39	2.39	2.39	0
6	2.38	2.37	2.375	0.01
7	2.37	2.36	2.365	0.01
8	2.38	2.36	2.37	0.02
9	2.41	2.4	2.405	0.01
10	2.37	2.37	2.37	0
11	2.35	2.36	2.355	0.01
12	2.36	2.37	2.365	0.01
13	2.38	2.38	2.38	0
14	2.37	2.36	2.365	0.01
15	2.38	2.37	2.375	0.01
16	2.36	2.35	2.355	0.01
17	2.35	2.36	2.355	0.01
18	2.37	2.37	2.37	0
19	2.36	2.34	2.35	0.02
20	2.38	2.36	2.37	0.02

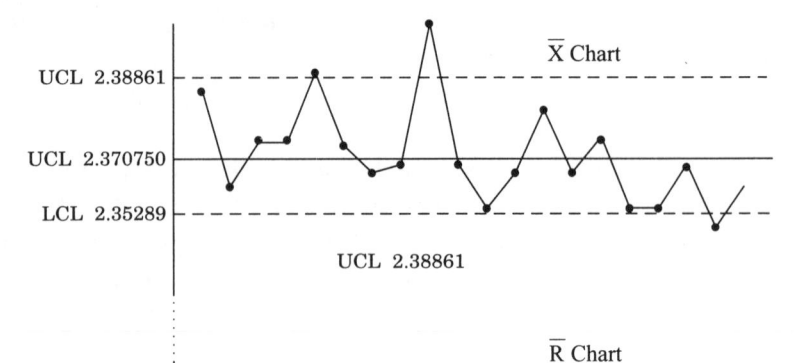

FIGURE 2.21: GAUGE CAPABILITY

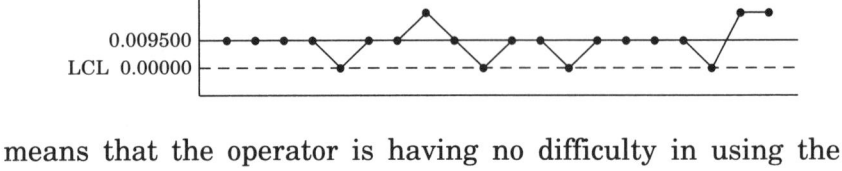

```
                                    X̄ Chart
UCL  2.38861  ─ ─ ─ ─ ─ ─ ─ ─ ─ ─ ─ ─ ─ ─ ─ ─ ─ ─

UCL  2.370750 ─────────────────────────────────

LCL  2.35289  ─ ─ ─ ─ ─ ─ ─ ─ ─ ─ ─ ─ ─ ─ ─ ─ ─ ─

              UCL 2.38861


                                    R̄ Chart
UCL  0.03104  ─ ─ ─ ─ ─ ─ ─ ─ ─ ─ ─ ─ ─ ─ ─ ─ ─ ─

     0.009500
LCL  0.00000  ─ ─ ─ ─ ─ ─ ─ ─ ─ ─ ─ ─ ─ ─ ─ ─ ─ ─
```

means that the operator is having no difficulty in using the gauge. The standard deviation of measurement error is calculated — using the formulae in Appendix A — as

$\sigma(\text{gauge}) = 0.0084$

The precision of the gauge is generally taken to be 6 times its σ and acceptable gauges should have a precision to tolerance ratio less than 10 per cent. This means that the variability inherent in the gauge is less than one-tenth the allowed variation within the product's tolerance specification. The precision of the gauge in our example is calculated as:

$6 \times \sigma = 0.0504$ mm

Measurements taken with this verniers will have a variability of +/– $3\sigma = 0.0252$ mm due to the measurement error within the gauge. The specification for the component being measured is

Upper specification = 2.65

Lower specification = 2.00

The ratio of gauge precision to specification tolerance is:

$$\frac{P}{T} = \frac{6\sigma}{USL - LSL} = \frac{0.0504}{0.65} = 0.775 = 7.75\%$$

Total Quality Management Tools

The result implies that the gauge precision is adequate for the product's specification. If the result was greater than 10 per cent, the engineer would need to investigate alternative means of measuring the product. Gauge capability studies ensure the right tools are used for each measurement task.

Statistical Process Control of Attribute Data

So far SPC has been described in terms of controlling measurement or quantitative data. It is also capable of controlling qualitative or attribute data such as pass/fail, defect/good, go/stop. The level of defects from the soldering process in our earlier example corresponded to attribute data. Statistical process control can be used to monitor the level of defects being generated and to highlight any trends in a similar way to that in which measurement data is controlled.

The reject level can be measured in a variety of ways. Two of the more common are presented here: the "P chart" and the "C chart". The type of chart used depends largely on what the manufacturing team are interested in controlling and improving. The first of these charts, the P chart, measures the proportion of defect units produced by the process.

FIGURE 2.22: ATTRIBUTE P CHART: MEASURES PROPORTION OF DEFECT UNITS WITHIN A SAMPLE

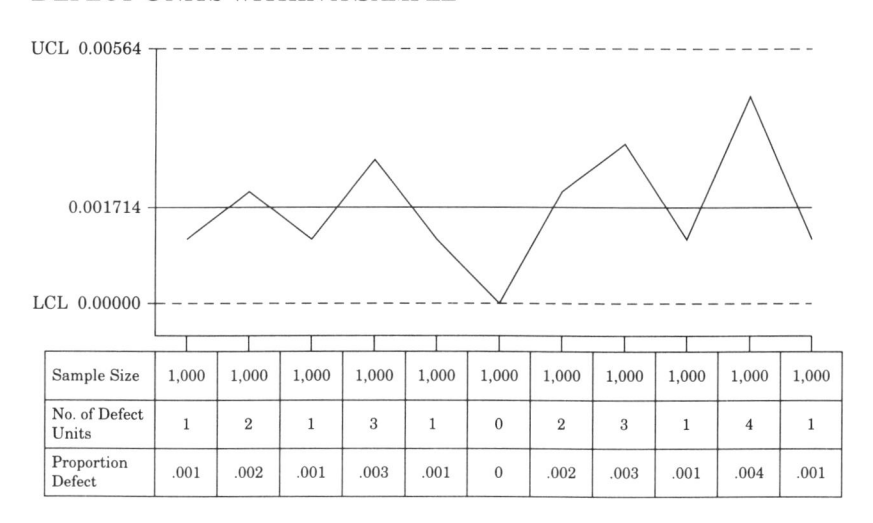

Sample Size	1,000	1,000	1,000	1,000	1,000	1,000	1,000	1,000	1,000	1,000	1,000
No. of Defect Units	1	2	1	3	1	0	2	3	1	4	1
Proportion Defect	.001	.002	.001	.003	.001	0	.002	.003	.001	.004	.001

UCL 0.00564
0.001714
LCL 0.00000

The proportion is calculated at regular intervals and noted on the control chart. The chart indicates the proportion of scrap generated by the process over time. Any out-of-control trends are highlighted for further investigation and resolution.

The proportion chart measures the number of units that were found to be non-conforming, but some units may be defective for several reasons. In the solderwave example, a unit may have a number of non-conformities; it may have broken tracks and simultaneously some components on the board may be unsoldered. The proportion chart logs this as one defect unit, while another form of attribute chart — a C chart — notes the number of non-conformities found. The C chart logs the number of flaws found in a constant sample size. The C chart emphasises the levels of flaws generated as opposed to the number of defect units. The formulae for calculating the upper and lower control limits are provided in Appendix A.

FIGURE 2.23: ATTRIBUTE C CHART: MEASURES NUMBER OF NON-CONFORMITIES WITHIN A SAMPLE IN A CONSTANT SAMPLE SIZE

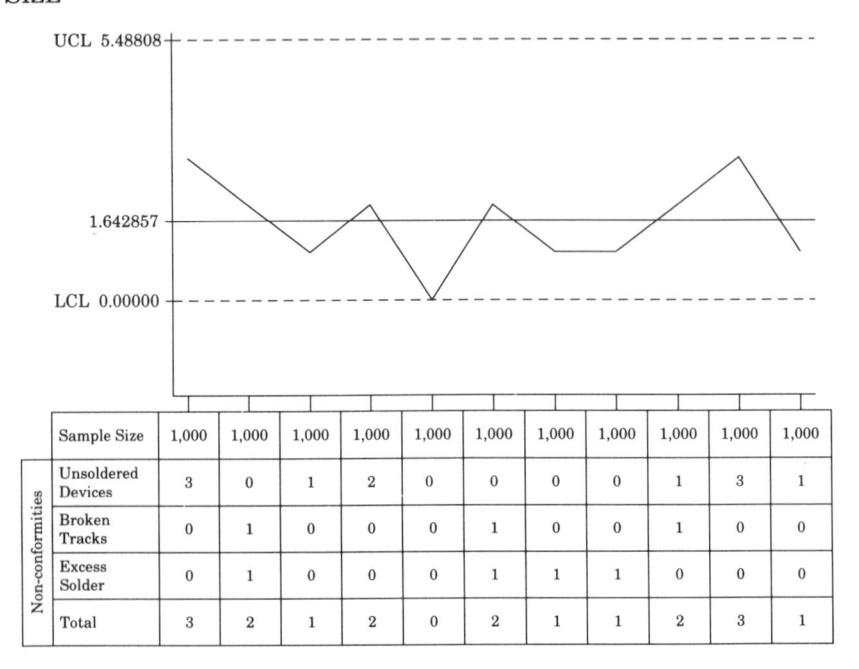

	Sample Size	1,000	1,000	1,000	1,000	1,000	1,000	1,000	1,000	1,000	1,000
Non-conformities	Unsoldered Devices	3	0	1	2	0	0	0	0	1	3	1
	Broken Tracks	0	1	0	0	0	1	0	0	1	0	0
	Excess Solder	0	1	0	0	0	1	1	1	0	0	0
	Total	3	2	1	2	0	2	1	1	2	3	1

Statistical process control establishes the current performance level of the process for attribute data. Any deviation in this performance level is highlighted in the same manner as with variable charts. The charts provide quantifiable data against which the process performance can be predicted, controlled and improved.

IMPLEMENTATION GUIDELINES

TQM is a company-wide approach to quality. It involves dispersing new problem-solving ideas throughout the organisation. Training is a key part of TQM. A broad spectrum of people need to learn and appreciate the power of the techniques available. Different companies approach this in different ways. One factory I worked in allocated three full days training for all employees to learn about the principles of TQM and the problem-solving techniques. Problem-solving tools were introduced and used on sample problems. The employees returned to their workplace armed with an understanding of improvement targets and methods for achieving improvement.

Translating the classroom onto the production floor is perhaps the hardest part of TQM. Sometimes middle management are sceptical of the benefits of TQM. Too often, training is provided but no opportunities arise in which to use it. The successful implementation of a pilot production line should enable a wide range of employees to understand the benefits of TQM on the factory floor. A more consistent process generally makes life easier for supervisors and equipment operators and therefore is a good advertisement for the TQM approach.

Another approach is to insist that middle managers have flow charts that describe their process, that performance gaps against targets are listed and that effort is clearly invested in eliminating the gap. This approach may force sceptical middle managers to adopt the TQM techniques on the factory floor in order to bridge performance gaps. Ideally, having experienced the improvement derived from the techniques, they will use the approach for a variety of problems.

A successful TQM programme should promote improvement work by advertising success stories throughout the organisation. The operational improvements gained should be communicated to all and the improvement team congratulated. Success stories help to promote the concept of TQM. Many organisations have internal newsletters that report on these success stories. Companies often arrange internal competitions for the best problem-solving team and the winning team presents their work to the entire factory and is rewarded for their effort. These activities help to disperse the ideas and techniques of TQM.

Improvements in metrics such as defects produced per million or the capability of the factory's process relate to the success of the TQM culture. Other important metrics include the level and type of customer complaints, the level of cost reduction or productivity levels. The direction of improvement work may have to be redirected from time to time to address each of these metrics.

TQM is a continuous improvement programme and management should therefore establish regular review meetings to monitor progress against targets. Once the momentum is established, it should be maintained through regular reviews and promotional activities.

TEAM DYNAMICS

TQM problem-solving techniques work best in a team environment where a cross-functional group of people come together with the purpose of eliminating a quality problem. Techniques like brainstorming, cause-and-effect analysis, failure modes and effects analysis, among others, are best used in a team environment where innovative solutions are generated. The concept of teamwork is alien to many people and many company cultures. Yet the ability to create and motivate a team is a critical part of the total quality management programme.

The old method is for managers to delegate problems to individuals, making them responsible for finding the solution. The engineer or supervisor attempts to find a solution by reorganising resources and systems under their control or by apportioning blame to other departments. Any implemented solution is limited by the narrow focus taken in problem solving. It is too easy to say that the engineering department does not understand the process, the quality department is too stringent or that operators do not care about their work. The real solutions to problem are found when proven problem-solving techniques are used in a cross-functional team environment.

For companies that don't have a team culture, the transition can be difficult and frustrating. Members of the team may not know what is expected of them and may perceive the whole exercise as ineffective and time wasting. It may provide some comfort to know that most teams experience a rocky start before settling down to the work at hand. Tuckman (1965) identified four transitional stages that groups go through:

- Forming
- Storming
- Norming
- Performing.

As the team develops, its members' level and quality of participation changes; in general, it is only when the team has transcended the Forming, Storming and Norming phases that it becomes effective. When the problem is of crisis proportions, and the seriousness evident to all members, transition through these phases is rapid. With lesser problems, it can take many sessions for the transition to happen.

Stage 1: Forming

At the forming stage, the team consists of individuals rather than team players. There is no loyalty to the group, as each person brings their own agenda to the meeting. Attendees may not be familiar with the team environment or working with people from other departments. Members will be assessing each other and making their positions and attitudes known.

This is a time for strong direction from the team leader. People will want to know what the team is expected to accomplish and how it plans to achieve its goals. Many companies prescribe mandatory problem-solving steps in the form of a questionnaire, which helps to focus the group effort. The questionnaire describes the purpose of the team and acts as a record of the team's work. Typical steps are:

1. Define the problem
2. List the potential root causes
3. Select the most likely causes
4. Verify each of the causes
5. Has the root cause been identified? If not, return to stages 1 and 2
6. Identify potential solutions
7. Verify effectiveness of solutions
8. Implement solutions.

Such a company-wide approach to solving problems avoids the need to re-establish ground rules at the initial stages of each team and therefore improves its effectiveness. All members understand that their function is to complete the questionnaire, and in so doing, to resolve the problem. The completed questionnaire can be filed and used by other areas of the business to solve similar problems.

Stage 2: Storming

Teams generally enter a storming phase, where conflict between the various team members can be very destructive. Conflict arises because team members have different functional business goals. The supervisors want the line to keep running, while the engineering staff want access to equipment. Conflict can arise out of personal animosity between group members. Internal politics means that different people bring their own personal agendas to the meeting. Conflict can stem from a variety of sources and the success of the team in surviving its storming phase will depend on the team leader's ability to handle conflict.

It is generally accepted that people adopt one of three approaches to conflict handling. Some people will constantly attempt to avoid conflict. These team members believe that "it is not nice to fight" and will stifle good ideas rather than jeopardise the harmony of the team. This is not an ideal situation, as good ideas are lost to the team. One approach is to encourage all ideas to be presented before decisions are taken — similar to the brainstorming techniques. In this way, people who avoid conflict are encouraged to participate to their full potential. The suitability of each suggestion is evaluated only after all ideas have been presented.

Some people have a combative style and could be categorised as tactless in their approach. Combative people often speak without thinking how other members perceive their analysis. They do not attempt to persuade; rather they challenge dissension from their opinions. Their approach often generates resentment and negativity towards the person and their ideas. This can lead to good ideas being dismissed because of where they originated rather than because of their inherent quality. Many of these people are emotional by nature and need help in seeing how their style creates a win–lose situation. They need to listen more and think before talking.

The third approach to conflict management is the collaborative approach, where the person seeks to find a win–win solution to conflict. Negotiation theory tells us to concentrate on what we have in common with those we are negotiating with and build on this common ground instead of focusing on differences. The collaborative style seeks to investigate the cause of conflict and to find a solution that is acceptable to both parties.

One of the easiest ways to minimise conflict is to use team consensus techniques, such as nominal group technique. Such techniques enable each idea to be given reasonable consideration and prioritised in accordance with perceived importance.

The nominal group technique for reaching consensus involves listing all ideas on a flipchart and asking each team member to write, on individual pieces of paper, a number against each idea, ranking its importance in descending order. In this way, the team member associates the number 1 with the most important idea, 2 with the second most important idea, and so on. The pieces of paper are gathered and the ranking for each idea is summed. The idea with the lowest score is the most favoured among team members.

For example, potential solutions to a production problem relating to product changeovers may be evaluated by a team of five people as:

Retrain operators	1 + 2 + 3 + 2 + 2 = 10
Upgrade equipment	2 + 1 + 1 + 3 + 1 = 8
Increase the minimum order quantity	3 + 3 + 2 + 1 + 3 = 12

In this example, the team perceives upgrading equipment as the most important of the three solutions presented. This technique minimises the risk of the team following the policy of the loudest members of the team and may also suit less forceful team members who avoid debate and conflict.

Stage 3: Norming

The norming stage involves developing the norms and patterns through which the team operates. How acceptable is it to arrive late for the meeting or unprepared? Some team members may test the cohesiveness of the team and experiment with the team to establish what will and will not be tolerated. This stage in team progress requires good skills from the chairperson to maintain acceptable standards within meetings. All team meetings should have a clear purpose that is easily understood. The team should recognise the route being taken to achieve the objectives. Each meeting should have a pre-prepared agenda and move the project closer to its concluding goal. The meetings should always begin on time, even if some members fail to arrive on time. Delaying meetings sends the wrong signals to the team. The chairperson should ensure discussion is focused and does not digress down "cul de sacs". Sideline discussions or meetings within meetings should be controlled and ideally avoided. The chairperson is responsible for ensuring that time is effectively used and not consumed on low priority issues. The meeting should conclude with a summary of decisions taken and actions to be taken. It should be clear who has responsibility for each action and the expected completion dates. The action list is only effective if the chairperson and team review progress on a regular basis. The meeting finishes on time and the chairperson verifies that all members can attend the next scheduled meeting.

Stage 4: Performing

When all members respect the norms of the group, the team enters the final phase of development where performance reaches optimum levels. The team develops a synergy which raises the confidence of the members to overcome obstacles that were previously considered insurmountable. Personal goals take a secondary place to the team's goals, and members feel a sense of belonging to a social entity.

A high performance team is an important asset for a company and they should be cultivated and encouraged to pursue their ideas. However, as the original problem is resolved, the activity level within the team will naturally fall away. The company has an opportunity to assign new problems to the group and maintain a high rate of improvement work.

FIGURE 2.24: TEAM PERFORMANCE CYCLE

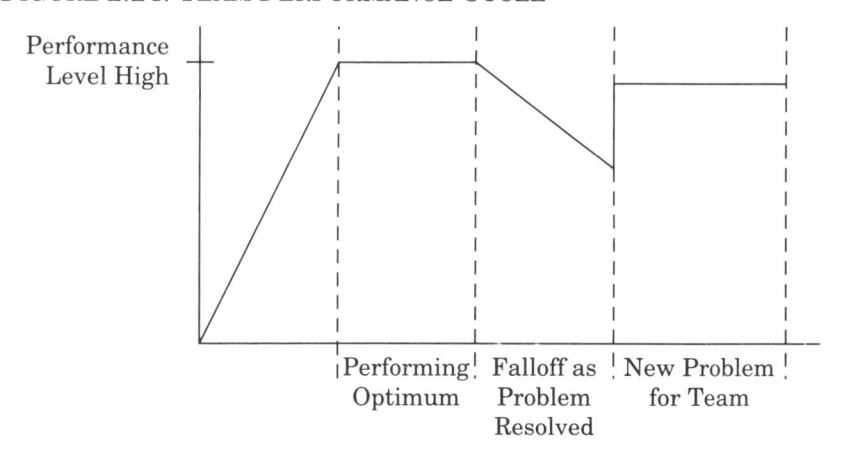

```
Performance
Level High
             Performing  Falloff as   New Problem
             Optimum     Problem      for Team
                         Resolved
```

As teamwork becomes a feature of the work environment within the company, the overall level of improvement work increases. More people spend a higher proportion of their day performing improvement tasks. This increase in innovative work can transform the stodgy company into a vibrant progressive organisation. Figure 2.25 shows how each person's type of work changes as the company develops a teamwork culture.

FIGURE 2.25: RATIO OF DAILY WORK TO IMPROVEMENT WORK

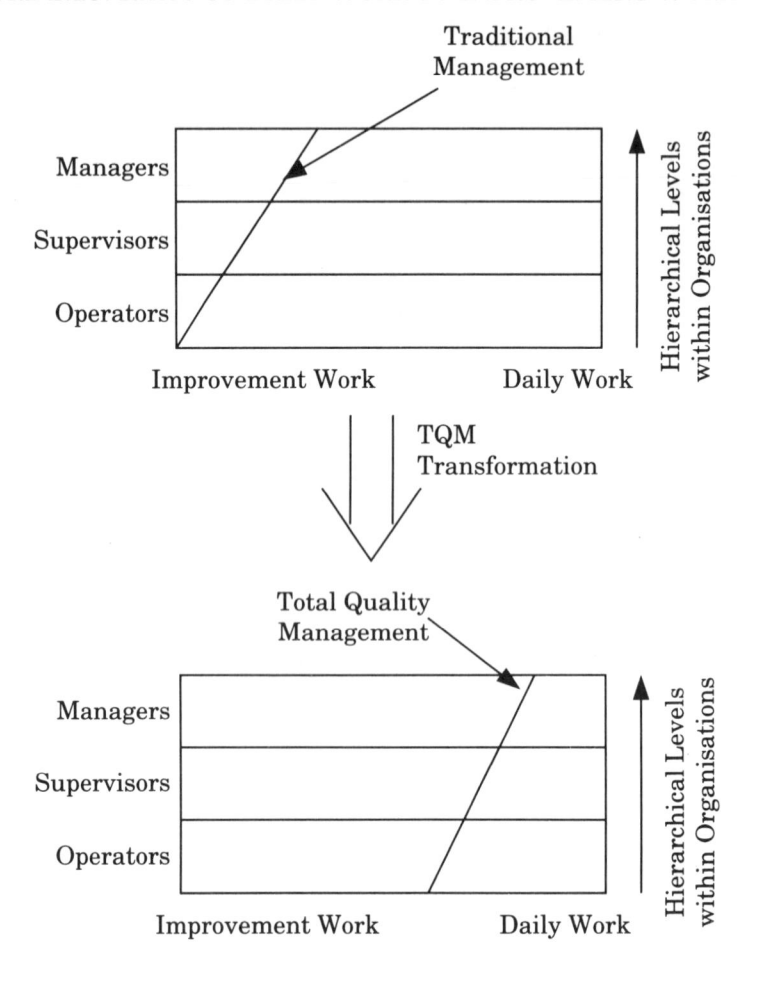

CONCLUSION

Quality is a competitive advantage that has been monopolised by Japanese firms in recent decades. Western companies are applying Deming's concepts in a desperate attempt to bridge the gap. Japanese companies have educated international companies by demonstrating there does not have to be a trade-off between product quality and cost. Continuous improvement in quality requires a serious study of the techniques available and a concerted effort to apply them within the workplace.

Deming recommended that continuous reduction of variation in all processes was the means to improving quality. Variation reduction establishes consistency in performance, which is the starting point for process improvement.

The tools for introducing consistency and process improvement have been described in this chapter. They include checksheets, brainstorming, cause-and-effect diagrams, failure modes and effects analysis, checklists, histograms, statistical process control, and process capability studies. Many other techniques exist which are covered by books listed in the bibliography. These tools help the improvement team to quantify the problem, to identify the causes of problems and to eliminate or control them.

Total quality management is a company-wide approach to quality improvement where all employees are trained in the use of quality techniques and are given the opportunity to partake in improvement teams. For many companies, it requires a culture change where teamwork and quantitative techniques replace subjectivity in improvement work.

TQM uses quality tools to quantify and analyse current performance against targets. TQM moves away from subjective interpretation of how well a person or function is performing. By quantifying performance, the rate of improvement can be measured and driven. By applying quality techniques throughout the organisation, the rate of improvement increases and the factory's capability improves.

Customers have adopted a zero tolerance for sub-standard goods and service. More than ever before, customers have a choice of supplier. Those suppliers who don't adopt an effective quality system are destined to continue to lose market share and competitiveness.

Chapter 3

QUALITY SYSTEMS: ISO 9000 AND QS 9000

INTRODUCTION

The quality of all manufactured goods is highly dependent on the quality of raw materials used in their production. It is hard, at times impossible, to convert poor, inconsistent quality materials into top class products. Sub-standard raw materials introduce variability into the process, which results in scrap units being generated and additional cost incurred. The problem is summarised in the old adage: "you can't make a silk purse out of a sow's ear!". Most companies recognise the imperative of having consistently high quality raw materials and place product quality ahead of price when making purchasing decisions. Purchasing personnel consider the overall cost of parts, including the cost of poor quality, when awarding contracts. Only quality-conscious suppliers are considered for tenders; any other approach could be described as "penny-wise but pound-foolish".

Historically, companies have protected themselves by defining quality requirements which their suppliers must adhere to. The requirements focus on ensuring that suppliers do not ship defective parts to their customers and that customer complaints are properly addressed. Suppliers have to put systems in place to protect the customer from defective parts and maintain records describing the results of inspections, improvement work, preventative procedures, and so on. The systems and records are regularly audited by customers to ensure compliance with the criteria. This approach to supplier quality attempts to ensure that only quality-conscious suppliers provide raw materials and the opportunity for defective parts to be shipped to the customer is minimised.

In the past, this strategy placed heavy burdens on suppliers who had to cope with a wide range of quality criteria from different customers. They were required to update their systems as each customer changed their criteria and, in addition, to facilitate customer audits of their premises. This led to resources being wasted in managing a proliferation of paperwork, especially as many of the requirements were common to the various customers. For this reason, amongst others, agencies like British Standards and, later, the International Standards Organisation (ISO) established a set of standards which defined product design, manufacturing, inspection and installation criteria. These criteria are audited by a third party and satisfy the expectations of many companies for their suppliers. As a result, many companies do not audit those suppliers registered for these internationally recognised standards.

The International Organisation for Standardisation developed the ISO 9000 series quality standard for the manufacturing and service industries in 1987, and it was immediately recognised by the industrialised countries. The European Union countries adopted the standard as EN 29000, and in the United States, it was adopted as ANSI/ASQC Q90. Companies registered to the ISO 9000 standard are complying with internationally recognised quality criteria.

There are three immediate benefits for companies adhering to the ISO 9000 quality standard. Firstly, adherence to the standard is a precondition for doing business with many companies. This measure of attainment enables the company to supply these corporations. Its widespread recognition means that many firms don't audit suppliers that are accredited to suppliers. For these companies, registration to the standard is satisfactory proof of an effective quality system.

The second benefit is the marketing potential derived from being accredited to an internationally recognised quality standard. The company can present itself as quality focused to potential new customers.

The third benefit is that adherence to a recognised quality standard provides an element of protection for companies who may face insurance claims from producing defect units. In both America and Europe, producers are responsible for the safety of their products in the hands of their consumers. The producer is liable for damage caused by a defect in the article. Quality assurance techniques are recommended by insurance companies as the best approach to avoiding exposure to excessive product liability claims. Adherence to the ISO 9000 standard provides objective evidence of top management's commitment to quality within their factory and therefore minimises exposure to claims of negligence.

However, the ISO 9000 series is generic by nature and aims to satisfy the needs of a wide range of industries. Companies like the big three automotive manufacturers and major truck manufacturers in the United States had more stringent requirements for their suppliers. In 1988, they combined their resources to develop common criteria. This led to the establishment of QS 9000 in August 1994. A supplier gains certification to QS 9000 by either second party audit by the automotive customer or a third party audit by accredited agencies like BSI.

QS 9000 contains the basic quality requirements of the three big automotive manufacturers and the main truck manufacturers. It aims to eliminate waste, prevent defects and install continuous improvement strategies to improve the capabilities of their supplier base.

The QS 9000 standard is made up of three sections. The first section relates to the ISO 9000 standard. Suppliers have to be registered to the ISO 9000 series of standards before being audited for QS 9000. The QS 9000 document details the 20 elements described by ISO 9001: 1994, but some of the elements are expanded to provide a clear interpretation of what the customer wants under the heading. In this way, much of the ambiguity in terminology of ISO 9000 is overcome with specific definition.

The second section details extra quality criteria that are common to Ford, General Motors, Chrysler and the major truck companies. This section describes requirements such as *Production Part Approval Process, Continuous Improvement*, and *Manufacturing Capabilities*.

The third section is customer-specific requirements. In this section each of the originators of the standard (Ford, General Motors, Chrysler and the truck manufacturers) specifies unique requirements that are applicable only to their particular suppliers. In order to be accredited to the QS 9000 standard, the supplier need only comply with those criteria in this section which are specified by their customer. The supplier needs to discuss this section with their customer.

Quality systems like ISO 9000 and QS 9000 should save money for both the supplier and the customer. From a supplier perspective, there is the initial cost of setting up the system to comply with the criteria, but the end result should be reduced defects generated, reduced resources invested in inspection and rework, and improved relationships with customers. From a customer perspective, the system should ensure that only units that are within specification are received. Resources are not wasted from working with substandard goods, and ideally, incoming inspection can be eliminated. The systems will not guarantee that only good units are produced, but if companies adopt the recommended quality system, reject levels should reduce and continue on a downward trend.

This chapter continues with a description of the ISO 9000 standard. An overview is given of the registration process and the criteria that have to be satisfied and implementation advice is provided on effectively satisfying the criteria. The extra requirements of QS 9000 are described and recommendations are made on registration to QS 9000.

DEFINITION OF A QUALITY SYSTEM

The ISO 9000 quality standard defines a quality system as "organisational structure, procedures, processes and resources needed to implement quality management". The same document defines quality management as "all activities of the overall management function that determine the quality policy, objectives and responsibilities, and implement them by means such as quality planning, quality control, quality assurance and quality improvement within the quality system". Management responsibility and the quality policy are the first requirements specified in the ISO 9000 standard.

SUITABILITY OF ISO 9000

A major advantage of a quality system like ISO 9000 is that it provides a well-documented system defining how tasks are performed for each person. It facilitates consistency in how tasks are performed. Internal and external audits ensure that the system stays alive and continuously improves the organisation's performance. Standards like ISO 9000 have a major marketing and corporate image value. Customers are impressed by companies who guarantee a high level of performance. ISO 9000 is respected by customers as a symbol of quality and as such can be a selling point for products.

ISO 9000 can have major implications from a product liability perspective. In cases where the customer takes the company to court regarding the quality of the product supplied, the company can be accused of negligence. Accreditation to ISO 9000 publicly demonstrates that the company is committed to quality and may help in minimising or avoiding claims for damages.

On the negative side, ISO 9000, by its nature, results in a lot of documentation. This requires much administrative work to achieve and maintain the standard. The level of administration depends on how the system is installed. A "keep it simple" philosophy is very important. Excessive documentation on how the company operates may stifle innovation because it does not accommodate change and, in general, is not required for compliance to the standard.

The significant benefits of the standard must be weighed against the administration costs and the registration fee payable to the government body responsible for regulating the standard. The level of documentation appropriate to a company depends on its circumstances. It is a balancing act between satisfying the needs of the standard and the company.

OVERVIEW OF THE ISO 9000 STANDARD

There are six documents making up the ISO 9000 series. These are ISO 8402, ISO 9004, ISO 9000, ISO 9001, ISO 9002 and ISO 9003. The first three documents — ISO 8402, ISO 9004 and ISO 9000 — define the vocabulary used in the standard and provide guidelines on how to implement it. The latter three documents describe the requirements of the standard as they apply to different aspects of the business.

The document ISO 9001 describes the quality criteria which must be fulfilled for businesses involved in product design, manufacturing and final inspection/test. The ISO 9002 document is a subset of ISO 9001, and covers only product manufacturing and final inspection/test. ISO 9003 is a subset of ISO 9002 and covers criteria for companies specialising in final inspection and test only.

Accreditation to the ISO 9000 standard is sought against the criteria specified in either ISO 9001, ISO 9002 or ISO 9003. The particular constituent most relevant will depend on the company's business area. Factories with R & D facilities will probably opt for accreditation to ISO 9001, whereas factories that produce according to a customer's design or industry standard may opt for ISO 9002. Factories which are mostly concerned with inspection of their end product will opt for ISO 9003.

The ISO 9001 document includes the criteria specified in both ISO 9002 and ISO 9003. Therefore by describing ISO 9001, the other two documents are explained by default.

Registration to ISO 9000

Companies can apply for registration (certification) to the ISO 9000 standard through a number of agencies. These agencies audit the company to verify compliance to the standard and maintain a registrar of all companies that they certify as conforming to the standard. There are several agencies available from which the company can choose to apply for registration. For example, in Ireland, there is the National Standards Association of Ireland (NSAI); in England, the British Standards Institution Quality Assurance (BSI); in France, Association

Française pour l'Assurance de la Qualité (AFAQ); and in Germany, the German Association for the Certification of Quality. The cost associated with registration varies between the agencies, but in general, fees are charged for:

- Application
- Pre-certification inspection
- Annual registration
- Annual surveillance.

The company seeking certification contacts the certifying agency, pays the relevant fees, and supplies a quality manual to the agency, describing how the company complies with the standard's criteria. Assuming that the quality manual is acceptable, the agency performs an on-site audit to verify that the company conforms to the standard. After successfully passing the audit, the company's name is added to a register maintained by the agency. The agency inspects the accredited company from time to time (in general, once or twice a year) to ensure the standard is maintained. If the standard criteria are consistently breached, the company's name may be removed from the register.

ISO 9001: 1994 Quality Requirements

There are 20 elements in the ISO 9001 standard. The paragraphs and sub-paragraphs are listed below and an overview is provided of their contents. The reader should read the actual standards before attempting to adopt the standard.

4.1 Management Responsibility
   4.1.1 Quality Policy
   4.1.2 Organisation
      4.1.2.1 Responsibility and Authority
      4.1.2.2 Verification Resources and Personnel
      4.1.2.3 Management Representative
   4.1.3 Management Review

4.2 Quality System
- *4.2.1 General*
- *4.2.2 Quality System Procedures*
- *4.2.3 Quality Planning*

4.3 Contract Review
- *4.3.1 General*
- *4.3.2 Review*
- *4.3.3 Amendment to a Contract*
- *4.3.4 Records*

4.4 Design Control
- *4.4.1 General*
- *4.4.2 Design and Development Planning*
- *4.4.3 Organisational and Technical Interfaces*
- *4.4.4 Design Input*
- *4.4.5 Design Output*
- *4.4.6 Design Review*
- *4.4.7 Design Verification*
- *4.4.8 Design Validation*
- *4.4.9 Design Changes*

4.5 Document and Data Control
- *4.5.1 General*
- *4.5.2 Document and Data Approval and Issue*
- *4.5.3 Document and Data Changes*

4.6 Purchasing
- *4.6.1 General*
- *4.6.2 Evaluation of Subcontractors*
- *4.6.3 Purchasing Data*
- *4.6.4 Verification of Purchased Product*
    - 4.6.4.1 Supplier Verification at Subcontractor's Premises
    - 4.6.4.2 Customer Verification of Subcontracted Product

4.7 Control of Customer Supplied Product
4.8 Product Identification and Traceability
4.9 Process Control
4.10 Inspection and Testing
- *4.10.1 General*

4.10.2 Receiving Inspection and Testing
4.10.3 In-Process Inspection and Testing
4.10.4 Final Inspection and Testing
4.10.5 Inspection and Test Records

4.11 Control of Inspection, Measuring and Test Equipment
4.11.1 General
4.11.2 Control Procedure

4.12 Inspection and Test Status

4.13 Control of Non-conforming Product
4.13.1 General
4.13.4 Review and Deposition of Non-Conforming Product

4.14 Corrective and Preventative Action
4.14.1 General
4.14.2 Corrective Action
4.14.3 Preventative Action

4.15 Handling, Storage, Packaging, Preservation and Delivery
4.15.1 General
4.15.2 Handling
4.15.3 Storage
4.15.4 Packaging
4.15.5 Preservation
4.15.6 Delivery

4.16 Control of Quality Records

4.17 Internal Quality Audits

4.18 Training

4.19 Servicing

4.20 Statistical Techniques
4.20.1 Identification of Need
4.20.2 Procedures

4.1: Management Responsibility

Management have to document their commitment to quality, how resources will be allocated and empowered to maintain the quality system. Management are expected to publicly declare their commitment to quality through a written quality statement. The statement declares their goals and focus on quality. They must be able to explain how their distribution of resources supports their declared focus on quality.

4.2: Quality System

The company must document how it ensures that its product conforms to specified requirements. It has to describe the steps it takes to ensure the product is built within the customer's specification. A quality manual has to be written describing the components of the quality system and how the various criteria are satisfied.

4.3: Contract Review

The company has to define how contracts are reviewed to ensure they are clearly specified and within the company's capability before undertaking the contract. The company must place the interests of its customers high on its priority list. New customer contracts should be reviewed at an early stage to ensure the company is capable of meeting all the targets set by the customer. The customer should be contacted where targets cannot be met.

4.4: Design Control

Procedures are required to control and verify the design of product so that the design brief and customer are satisfied. The company has to ensure its understanding of the design brief corresponds to the customer's perception. All necessary questions regarding the product's fitness for its end purpose should be clarified at the initial phase of the project.

4.5: Document and Data Control

A system for controlling and storing all documents and relevant data has to be set up. ISO 9000 results in a multitude of procedures, product specifications, audit reports, calibration reports, etc. Documents must be reviewed by nominated personnel before becoming official and enforceable within the company. They have to be properly filed, maintained and utilised.

4.6: Purchasing

The company has to document how it ensures that purchased product always conforms to specification. Quality begins with the manufacturer's suppliers. Consistently good quality raw materials are a major step towards producing quality products.

4.7: Control of Customer Supplied Product

In cases where the customer supplies raw material to be incorporated into the end product, the company must establish procedures to verify and control the quality level of supplies. In this situation, the company has been entrusted with the responsibilities for handling, storing, and manipulating their customer's goods. Procedures are needed to describe how the company plans to ensure these goods are maintained in top quality.

4.8: Product Identification and Traceability

When appropriate, the company must document its capability to identify and trace the product route through the different stages of the process. Information, such as which machines produced the product and to what production batch does the unit belong, is helpful in resolving process quality problems. This information is also critical when it is necessary to recall potentially defective product from the marketplace.

4.9: Process Control

Work instructions are required to define how tasks are correctly carried out at each stage of the process. The instruction procedures provide a step-by-step guide to performing the work correctly. Product verification techniques that ensure good quality must be incorporated into this documentation.

4.10: Inspection and Testing

Inspection and test is performed on incoming goods and on product at different stages of the process. The criteria and method used when performing the inspection and tests need to be documented. This requirement ensures that the product is within specification at different stages of production.

4.11: Control of Inspection, Measuring and Test Equipment

Equipment used to perform product inspection and testing needs to be calibrated to ensure its accuracy. The inspection task can only be as good as the accuracy inherent in the inspection tools. The accuracy of these tools has to be periodically checked and traceable to national standards.

4.12: Inspection and Test Status

The test status of product in an operation must be easily recognisable. A method has to be defined which clarifies if product is awaiting test or has been tested. It is all too frequent that product awaiting test is in close proximity to product that has been tested as good and product that has been tested as reject. If this product becomes mixed up, the customer could easily get either untested product or even reject product. Wherever product is within the factory, its status should be easily recognisable.

4.13: Control of Non-conforming Product

Product that has been identified as not conforming to documented specifications must be segregated from other product. Non-conforming product must be controlled in such a way as to ensure it is not mistakenly used or installed, etc.

4.14: Corrective and Preventative Action

The causes of non-conforming product have to be documented and plans put in place to prevent their recurrence. Records of the occurrence and reaction to non-conforming product must be maintained. Records that trace the product route through the process often yield valuable information regarding the sources of defects. The company must pay special attention to customer complaints, demonstrating how preventative action is taken to avoid recurrence.

Quality Systems: ISO 9000 and QS 9000 101

4.15: Handling, Storage, Packaging, Preservation and Delivery

Procedures are required which document how product quality is protected and preserved throughout the order fulfilment process. These instructions are often incorporated into the various work instruction procedures.

4.16: Control of Quality Records

Records provide objective evidence that the standard is being adhered to. Records of inspection results must be maintained and available to the auditor upon request. Records must also be maintained regarding the operation of the quality system — for example, the results of internal audits, customer complaints, etc.

4.17: Internal Quality Audits

The quality system must be monitored by internal staff on a regular basis to ensure conformance to the standard. Generally the company nominates personnel to perform internal audits to supplement those performed by the certification body. These audits provide an early warning of deviations from the written system.

4.18: Training

The company should evaluate each task to establish relevant training required. Records of employee training have to be maintained. The aim is to avoid having untrained personnel putting quality at risk.

4.19: Servicing

When appropriate, procedures are required which describe how to perform, verify and report servicing and installation work. This is important for companies like machine manufacturers who install their machines as part of the contract.

4.20: Statistical Techniques

Where appropriate, companies must establish statistical techniques for controlling and verifying process capabilities and product characteristics. Statistics are a powerful tool when correctly implemented, as described in an earlier chapter of this book.

DOCUMENTATION

Documentation of the quality system typically consists of four tiers:

- The Quality Manual
- Quality System Procedures
- Work Instructions
- Quality Records.

Quality Manual

The ISO 9000 standard requires that a quality manual be written, describing how the company has addressed each of the requirements of the standard. The manual should provide a concise description of the practices and procedures that constitute the quality system and should contain enough information to assure the auditor that the system adheres to the standard. It is usually submitted to the third party registrars before the certification process begins.

Generic quality manuals which can be customised to cater for the company's individual requirements are available off the shelf. Some companies offer generic manuals in software format so that users need only fill in the blanks to create their own manual. There is a danger in adapting these generic manuals as the end result may not reflect the peculiarities inherent within the company. Writing a manual from the beginning is an opportunity to describe the company's individual approach to achieving quality and satisfying the ISO 9000 standard. Having policy statements pre-written for the company may appear an empty gesture to company employees and the standard auditors.

Quality manuals tend to include:

- The company's policy statement where the senior company manager (or divisional manager seeking accreditation for their division) states the company's commitment to quality and the policy.

- Introduction to the company describing the business it operates, how the organisation is structured, how company processes operate (process flow diagrams), how responsibilities are distributed.

- Clarification of department personnel with responsibility and authority to develop, document and enforce a quality system that supports the ISO 9000 standard and the company's policy statement.

- Description of the documentation used to support the standard, such as work instructions, quality procedures and records.

The quality manual is the main document that describes the company's quality system. It is a requirement of the standard to have a quality manual but there is flexibility in the style and format of the manual — as long as it achieves its goal. A common structure is to lay out the manual sections in a sequence related to that in which the quality criteria are listed in the standard. Such a layout consists of:

- Title
- Table of Contents
- Foreword (0.0)
- Revisions (1.0)
- Distribution List (2.0)
- Introduction and Scope of Manual (3.0)
- Control of Manual
- Management Responsibility (4.1): *Describe top management's commitment to quality in general terms and how responsibility and authority is delegated to ensure quality.*

- Quality Policy (4.1.1): *Define the company's objectives for and commitment to quality, describing how quality activities have the full support of top management. This policy statement must be signed by the most senior company manager.*
- Organisation (4.1.2): *Provide an organisation chart for the company portraying how the various company departments interact.*
- Responsibility and Authority (4.1.2.1)
- Statistical Techniques (4.2.0).

Each of these sections is detailed to reflect the company's compliance to the standard.

The above structure is a suggested format. The italics represent suggested information for inclusion in the manual. Various procedures are referenced throughout the manual to direct the reader to further sources of information on how the company adheres to the standard.

Structuring the Documentation System

Tier-two, -three and -four documentation combine to provide a detailed description of the quality system, its methods, practices and results. Tier-two documentation — quality system procedures — describes practices associated with the running of the department, or tasks not directly linked to any individual job. It may provide instructions that apply to a range of users. Instructions on how to perform an internal audit would apply to anyone undertaking such a task.

Tier-three documentation — work instruction — relates to a particular job description; for example, operating equipment. They provide step-by-step instructions on how to perform tasks which affect the quality of the product produced.

Tier-four documentation — records and forms — are used for collecting information. Internal auditing checklists are used to record results from internal audits; calibration records are proof that test and measurement equipment is calibrated. These and other records have to be maintained and made available to auditees on request.

The implementation team should spend time considering how to structure the documentation system. For example, according to the standard, the company must have a system for evaluating the performance of its subcontractors. If no such system exists, the implementation team has to decide on performance metrics against which sub-contractors will be measured. These metrics need to be agreed with the subcontractor and a method established for monitoring performance. The ISO 9000 standard requires that this new system be documented.

The team must decide if the new system should be documented as a work instruction procedure for the person in charge of receiving or if it should be documented as a quality system procedure which is referenced in the work instructions of the person responsible for receiving goods. The team must also develop forms on which to gather information and a system for analysing the results. Should the forms be part of the work instruction procedure or the quality system procedure? The records of performance will have to be analysed against acceptance limits and maintained as a record of compliance to the ISO 9000 criteria.

Tier-Two Documentation: Quality System Procedures

In general, quality system procedures can be developed by expanding on the information given in the quality manual for each of the standard's requirements. The quality system procedure describes, in simple language, a step-by-step instruction on how to conform to the specific requirement.

Quality system procedures describe who performs the various tasks within the organisation, how to establish quality plans for the introduction of new products, how to make changes to documentation, how inventory is tracked, etc. Very often, quality system procedures relate to a range of people while work instruction procedures are specific to a particular job description.

Tier-Three Documentation: Work Instructions

Work instructions describe how to perform various tasks such as equipment operation, product inspection at an inspection station or reworking.

Work instructions should be written in clear, easy-to-read language. It is important that the person performing the task is able to understand what the procedure states. The level of detail in work instructions varies. Exact instructions are usually required for simple tasks. Work instructions for more complex tasks may take a format similar to guidelines. The standard specifies that the appropriateness of the information depends on the particular circumstance of the company.

Tier-Four Documentation: Records

Records need to be maintained to provide evidence of compliance to the written procedures and to the standard. Records have to be maintained of inspection results, calibration results, customer complaints, corrective actions taken, training, etc. The team should systematically identify the records that need to be maintained for each element of the standard.

THE AUDIT

Compliance to the ISO 9000 standard is established by means of an audit. The purpose of the audit is to verify that documentation exists describing how each of the criteria is satisfied, and to verify that the standard is adhered to in practice.

The quality audit is defined in ISO 8402: 1994 as:

> A systematic and independent examination to determine whether quality activities and related results comply with planned arrangements and whether these arrangements are implemented effectively and are suitable to achieve objectives.

The ISO 9000 audit is performed by a third party auditor or team of auditors representing the registrar. The ISO 9000 auditing process consists of two parts: a systems audit and a compliance audit.

1. The *systems audit* is an inspection of the quality manual and associated procedures to verify that the systems have been defined to comply with the ISO 9000 criteria.

2. The *compliance audit* ensures that the documented system is carried out in practice and is effective and suitable to its objectives.

The company generally has to supply the quality manual and related procedures to the auditing team, typically, up to two weeks before the audit begins. This enables the auditors to ensure that the company has satisfied the basic requirements of ISO 9000 by documenting a quality system that complies with the standard. Any discrepancies in the documented system should be resolved before commencing with the on-site audit.

The on-site audit aims to gather objective evidence of compliance to the standard and also to highlight any discrepancies. Non-conformances are divided into major and minor. A major non-conformance would be the complete failure to address an element or paragraph of the standard. Another major non-conformance would be the failure to implement a procedure that clearly results in poor quality product being produced. Minor non-conformances would be criteria that are only partially satisfied but which do not put product quality in immediate danger. For example, training records that are available but are not completely filled in. The existence of a major discrepancy or too many minor discrepancies results in the company not being certified to ISO 9000.

There are three main components of the on-site audit: the opening meeting, the audit of the various company functions, and the exit meeting. The opening meeting establishes the format and objectives of the audit. It is an opportunity for both the audit team and company management to become acquainted. The scope of the audit is explained, the standard against which the facility is audited confirmed, the schedule for the audit established, and the evaluation procedure explained.

Guides must be provided for auditors to inspect the operations of the company. The guides should be familiar with the various functions being audited and should be able to explain various happenings that may be observed. For example, if the auditor notices product is being reworked, the guide should be able to take the auditor through the procedural steps that led to the decision to rework the product. The guide should be able to introduce the auditor to the various people responsible for the various activities being performed.

Notes are recorded of all observations relating to quality and the ISO 9000 standard. A checklist is used throughout to systematically verify that all of the standard's requirements are satisfied.

The audit concludes when a formal exit meeting is held at which a written report is presented describing the audit's outcome. All discrepancies should be resolved before the company is certified as ISO 9000 compatible. A corrective action plan is agreed and, depending on the level of discrepancies, a follow-up audit takes place and/or the company provides documentary evidence of resolution of discrepancies. Accreditation to the standard means the company's name is added to a register of accredited companies. ISO 10011-1: 1990 (E). Guidelines for auditing quality systems provides further information on how the auditing process should operate.

IMPLEMENTATION

The first stage of the implementation process is to establish a cross-functional implementation team and a management representative who will take responsibility for directing the effort. Team members should be chosen from all departments within the organisation, because ISO 9000 affects all departments.

The implementation team has to develop a clear understanding of what requirements are specified within ISO 9000. The team should study each of the elements in order to:

1. Clarify the requirements of each section. For example, in ISO 9001, "4.4.2.1: Activity Assignment" is a subheading under the element entitled "4.4: Design Control". The activity assignment states that: "The design and verification activities shall be planned and assigned to qualified personnel equipped with adequate resources." This single sentence can lead to a series of questions:

- What are the design and verification activities?
- How will they be documented?
- What level of skills will the person responsible for design and verification require?
- What level of resources will be required?

The team needs to examine each paragraph and subparagraph and understand the underlying requirements associated with each of them.

2. Examine the standard for links between the various elements. For example, the requirement stated above as "4.4.2.1: Activity Assignment" is closely related to the first sentence of "4.4.5: Design Verification", which states, "The supplier shall plan, establish, document and assign to competent personnel functions for verifying the design". Links also occur in other ways. The reference to "qualified personnel" in subparagraph 4.4.2.1 above, and the reference to "competent" in 4.4.5 is linked in part to paragraph 4.18, "Training": "The supplier shall establish and maintain procedures for identifying the training needs and provide for the training of all personnel performing activities affecting quality".

3. List all questions posed by the standard. This allows the team to evaluate the current quality system to establish the gap between the existing systems and the requirements of the standard. Where appropriate, the existing procedures may be adapted to satisfy the criteria.

4. Structure the documentation system. The team should decide the type of information to be contained in each of the tiers of documentation: the level of detail contained in the quality manual; the quality system procedures; the work instructions and record forms. Outlining the interface between these documents will enable the team to begin writing the various documents concurrently while reducing fear of duplicating information unnecessarily.

5. Document the new quality system and train employees. The team should list all procedures that need to be developed, who will be responsible for their development and who will need to be trained regarding their contents.

Once the system has been documented, a programme of internal audits must be established and undertaken to ensure compliance with the procedures in practice. The results of these audits are retained and regular meetings held to address any non-conformances discovered.

Implementation periods of approximately one year are typical for companies attempting to comply with ISO 9001. This time may be longer or shorter, depending on the appropriateness of the systems currently in place and the time and effort invested in the implementation.

THE QS 9000 STANDARD

The QS 9000 standard is more stringent in its requirements than ISO 9000. It supports its requirements with reference manuals that describe how quality plans should be developed and how the various processes should be controlled. It also contains separate sections that describe a part approval process, and reinforce the need for a continuous improvement philosophy to prevail within the organisation and the need to have a capable manufacturing process.

Suitability of QS 9000

Compliance to the QS 9000 standard is compulsory for all direct suppliers to Ford, General Motors or Chrysler by the end of 1997. Compliance to the standard enables their suppliers to ship direct to their customer's production line. It means that quality inspection of incoming goods is no longer required.

Part of the requirement of the standard is that suppliers to the automotive manufacturers should develop the quality systems of their subcontractors. Therefore, the requirements laid down in the standard may be enforced on subcontractors to these suppliers.

OVERVIEW OF QS 9000

The following provides an overview of the QS 9000 standard. It is not a substitute for a proper examination of the standard, but an introduction to the criteria specified. The standard consists of three sections. Section 1 is an expansion on the ISO 9000 standard. All existing 20 elements and subparagraphs of ISO 9000 apply within the QS 9000 standard. However, additional requirements are added. A summary of the extra requirements is given below. Section 2 specifies that suppliers must comply with their customer's Part Production Approval system and must demonstrate continuous improvement and manufacturing capability. Section 3 describes the individual requirements of each of the companies who originated the standard.

QS 9000 Section 1

The following is a summary of the additional requirements of QS 9000 beyond those specified by ISO 9000. The extra requirements take the form of additions to the existing ISO 9000 paragraphs and the addition of new paragraphs. The automotive industry refers to three manuals throughout the standard.

- *The Advanced Product Quality Planning and Control Plan (APQP):* The APQP manual covers market research, business strategy, product design and development, new product introduction and process improvement.

- *The Production Part Approval Plan:* Describes the documentation that must accompany any new part or proposed significant changes to existing parts before they are accepted by the customer.

- *Potential Failure Modes and Effects Analysis:* Describes how to perform a failure modes and effects analysis.

- *Statistical Process Control:* Describes the concepts involved in statistical process control and how it can be effectively used.

- *Measurement System Analysis:* Describes how to determine the suitability of inspection and measurement systems by performing repeatability and reproducibility studies.

Each of these reference manuals supports the standard by describing how to approve a part or process, how to set up statistical process control and failure modes and effects analyses and how to perform measurement capability studies. The standard refers to the relevant manual when appropriate to the criteria described. The supplier will have to satisfy the criteria in accordance with the relevant manual. Only those paragraphs with extra requirements and additional paragraphs are described here.

4.1.2.3: Management Review (extra requirement)

QS 9000 specifies that suppliers should use the Advanced Quality Planning and Control Plan. The supplier must use cross-functional teams to develop, maintain and review the quality system. QS 9000 recommends that teams should comprise representatives from sections such as marketing, engineering, materials, management information systems, subcontractors and customers, as appropriate.

4.1.4: Business Plan (additional paragraph)

QS 9000 specifies that suppliers shall document a business plan that evaluates market trends, benchmarks performance against competitors and plans the company's goals for the short- and long-term perspective. The supplier must have a process for collecting relevant business data and regularly updating the business plan. The supplier is encouraged to empower their employees to enable them achieve the plan's goals.

4.1.5: Analysis and Use of Company-level Data (additional paragraph)

The supplier must maintain records relating to performance and compare trends in productivity, quality, efficiency, etc., against competitors. Appropriate action should be taken, as

Quality Systems: ISO 9000 and QS 9000

necessary, to ensure the trends indicate progress towards achieving business goals.

4.1.6: Customer Satisfaction (additional paragraph)

The supplier must have a system in place for measuring customer satisfaction. The system should provide quantifiable, objective information relating to customer satisfaction. The level of customer satisfaction should be benchmarked against competitors and monitored by top management.

4.2.3: Quality Planning (extra requirements)

- *Special Characteristics:* The QS 9000 standard reiterates the need to use the Advanced Product Quality Planning and Control Plan. During the planning phase, special characteristics of the product shall be identified and controlled in the manner specified by the APQP manual and by each customer's individual requirements. Those characteristics which are considered special vary between Ford, General Motors and Chrysler. In general, any product or process characteristics that affect the product safety in use or relate to compliance with government regulations are considered to be special characteristics.

- *Cross-functional Teams:* A cross-functional approach is to be taken in identifying special characteristics. The team should be used to develop the necessary FMEAs, control plans and the improvement plans to minimise the risk priority number of FMEAs.

- *Feasibility Review:* The supplier must undertake a feasibility review that verifies its capability to manufacture the product to the correct quality standard in the correct volume. The feasibility review has to be documented in accordance with the APQP manual's section on Team Feasibility Commitment.

- *Process Failure Mode and Effects Analysis:* Process failure modes and effects analysis is a system of evaluating the effect of a failure mode, its causes and the measures installed to detect or prevent its occurrence. All special characteristics have to be evaluated using FMEA. Defect prevention must be emphasised instead of defect detection.

- *Control Plan:* A control plan documents the systems and controls used to ensure that only good product is produced. It is derived form the advanced product quality planning process. The steps identified by the cross-functional team as the means for ensuring quality are set up and documented as the control plan. Control plans should exist for each system and subsystem, as necessary, to ensure quality. The control plan must cover prototype building, pre-launch and production stages. The prototype and pre-launch control plans must include dimensional measurements, material and performance tests to be carried out. The production control plan provides a comprehensive description of the product and process characteristics, process control parameters, test specifications and measurement system. The standard stresses that the control plan is a living document and must be updated to reflect changes in the process and/or product.

4.4.2: Design and Development Planning (extra requirement)

QS 9000 lists the skills in which the design department should be appropriately competent. These skills include quality function deployment, value engineering, design of experiment, finite element analysis, failure modes and effects analysis, geometric dimensioning and tolerancing, solid modelling, reliability engineering, computer aided design and engineering, design for manufacture and assembly, and computer simulation.

4.4.4: Design Input (extra requirement)

QS 9000 specifies that the supplier should have appropriate resources and skills in computer aided design and engineering. Where these services are subcontracted, the supplier should provide technical leadership. The customer may waive the need for computer aided design capabilities.

4.4.5: Design Output (extra requirement)

The supplier must use techniques such as cost analysis, geometric dimensioning and tolerancing, FMEAs, etc., to ensure that the process is optimised and waste is minimised.

4.4.7: Design Verification (extra requirement)

The supplier must have a comprehensive prototyping system unless this need is waived by the customer or the product is a standardised item. The prototyping process must include product life testing and subcontractors for tooling, and raw materials used during prototype development must also be used during the mass production of the product. All test and performance data must be recorded and analysed in a timely manner to ensure the project is completed on time. Even when aspects of the prototyping are subcontracted, the supplier must provide technical leadership.

4.4.9: Design Changes (extra requirement)

QS 9000 requires that all design changes, even those instigated by subcontractors, must be approved by the customer before being implemented. The customer may waive this requirement. The manual on production part approval process describes the documentation that should accompany any request for change. With proprietary designs, the customer must be consulted regarding any changes to form, fit, function or performance.

4.5: Document and Data Control

4.5.1: General (extra requirement)

The supplier must have all documentation referring to a task available at the workstation where that task is performed. Such documentation includes work instructions, engineering drawings, inspection and test procedures, etc. The documentation must identify special characteristics using either the symbols specified by each customer in appendix C of the standard or by using equivalent in-house notation.

4.5.2: Document and data approval and issue (extra requirement)

The supplier must develop a procedure for the timely review and implementation of customer engineering standards/ specifications and changes to these. The supplier must record when the changes were implemented. All related documents must be updated as part of the implementation.

4.6: Purchasing

4.6.1: General (extra requirement)

The supplier must purchase their supplies from the customer's approved vendor list, when suitable vendors are listed. Any additional subcontractors have to be added to the list before becoming acceptable vendors. Materials used in the manufacturing of parts must comply with government regulations in the country of manufacture and the country of sale.

4.6.2: Evaluation of Subcontractors (extra requirement)

This section requires that the supplier should develop their subcontractor's quality system according to sections 1 and 2 of the QS 9000 standard. The subcontractor may be audited by the customer, by a customer-approved original equipment manufacturer or by an accredited third party.

The supplier must monitor the performance of subcontractors. They must demand 100 per cent delivery reliability from subcontractors. The supplier needs to provide adequate information such as production forecasts to enable the subcontractor to achieve 100 per cent on-time delivery.

4.6.3: Purchasing Data (extra requirement)

The supplier has to have a process for ensuring that subcontractors adhere to government regulations when producing parts.

4.9: Process Control (extra requirement)

The standard specifies that suppliers must comply with government safety and environmental regulations. The supplier should possess certificates proving compliance.

The supplier has to document the steps taken to comply with the customer's requirements, including the designation and control of special characteristics. The supplier should be able to provide documentation demonstrating compliance to the customer.

The standard points out that *all* product and process characteristics should be controlled, not just the designated special characteristics. However, extra attention is needed to ensure tight control of special characteristics.

QS 9000 specifies that the supplier must identify key process equipment and develop a total preventative maintenance programme for these machines. This should include a procedure documenting preventative maintenance activities, predictive maintenance activities, manufacturer recommended activities, etc. The levels of uptime should be monitored and improvement action taken as appropriate.

4.9.1: Process monitoring and operator instructions (*additional paragraph*)

The supplier is required to supply operators with detailed instructions on monitoring the process and operating the equipment. The instructions should be derived from work done in quality planning according to the Advanced Product Quality and Control Plan. The operator instructions make reference to part number and name, name of the operation, current engineering level date, required equipment, engineering and manufacturing standards, materials identification and disposition, statistical process control requirements, inspecting and test requirements, corrective action, set up instructions and tool change frequency and visual aids, if appropriate.

4.9.2: Preliminary process capability requirements (*additional paragraph*)

This section requires that a preliminary process capability study be performed for all processes associated with special characteristics. A target of 1.67 should be achieved within 30 working days unless otherwise specified by the customer. The preliminary process capability study must be reviewed with the customer at various stages of quality planning. Unacceptable preliminary process capability performance must be rectified by the use of mistake proofing and continuous improvement techniques. Attribute data from production runs should be used to drive continuous improvement efforts.

4.9.3: Ongoing process performance requirements (*additional paragraph*)

The customer may define ongoing process performance requirements. In the absence of specific requirements, the default is that process capability studies must be above 1.33 for stable processes and above 1.67 for chronically unstable but predictable processes. For non-normal, unpredictable processes, other techniques such as part-per-million defect are used to define minimum acceptable performance requirements. All significant events such as tool change, machine repair, etc. should be documented on the control charts. The supplier must document how they plan to cope with non-capable processes, perform 100 per cent inspection of product and improve the process so that it is capable. The supplier must install a continuous improvement programme, even for those processes which are proven capable.

4.9.4: Modified preliminary or ongoing capability requirements (*additional paragraph*)

When the customer has higher or lower requirements for capability of processes, the supplier should modify the control plan to reflect that requirement.

4.9.5: Verification of job set-up (*additional paragraph*)

QS 9000 requires that all job set-ups be verified by inspection of parts to ensure that they are within specification. Statistical techniques are recommended for this task. The standard also recommends that the first part off should be compared to the final part as a measure of stability of the process.

4.9.6: Process changes (*additional paragraph*)

The production part approval process applies to a part number and a process operating at a specific location, at a specified engineering change level, using specific raw materials, within a specific production processing environment. In general, any change to the part number or any of the above items requires prior approval from the customer.

4.9.7: Appearance items (*additional paragraph*)

Parts which have been customer-designated as "appearance items" must be evaluated by qualified persons with appropriate lighting and with access to masters which define the appearance standard.

4.10: Inspection and Testing

4.10.1: General (*extra requirement*)

Sampling plans for attribute data are to have acceptance criteria of zero defects. Acceptance criteria is to be documented and agreed with the customer. Suppliers are to use accredited laboratory facilities as requested by the customer.

4.10.2.3: Incoming product inspection (*extra requirement*)

The supplier has to use one or more of the following techniques as part of its incoming quality inspection:

- Statistical data
- Receiving inspection
- Quality audit of the subcontractor's quality system
- Evaluation of the part by an accredited laboratory

- Certificate of conformance by the subcontractor.

4.10.4: Final inspection and testing (*extra requirement*)

The supplier is expected to perform a layout inspection and functional verification according to customer-specific frequency and criteria laid out in section 3 of the QS 9000 standard.

4.11.3: Inspection measurement and test equipment records (*additional paragraph*)

Records must be maintained for all gauges, test, and measurement equipment describing calibration and verification results. The customer must be notified if suspect material has been shipped.

4.11.4: Measurement system analysis (*additional paragraph*)

All gauges and measurement equipment included in the control plan must be evaluated in accordance with the measurement system analysis manual. The analytical methods used should determine the repeatability and reproducibility of the measurement equipment used.

4.12: Inspection and Test Status (extra requirement)

Product that fails to pass inspection or is suspected of failing must be tagged as such and placed in a designated area. There should be no possibility of confusion between passed product and failed product. When the customer requires *supplemental verification*, such as early launch controls, such verification must be provided by the supplier.

4.13.4: Engineering Approved Product Authorisation (additional paragraph)

Any changes to the current approved process or part must have agreement in writing from the customer. The supplier must maintain records of agreement from the customer (including expiration date and quantity ordered). The authorisation must be identified on the shipment packaging. Similarly, the supplier must have agreement from the customer before accepting any change to its subcontractors' parts or process. Once the authorisation expires, the supplier must comply with the current customer specification.

4.14: Corrective and Preventative Action

4.14.1: General

QS 9000 requires that disciplined methods be used to solve problems related to internal or external non-conformances. The supplier must respond to non-conformances in a manner specified by the customer.

4.14.2: Corrective Action (extra requirement)

The supplier must analyse parts returned as defective from the customer. Records must be maintained of the finding and corrective action taken to prevent recurrence.

4.15.3: Storage

The supplier must establish a system for reducing inventory levels, maximising inventory turns and optimising the use of inventory.

4.15.4: Packaging

Generally, the customer has unique requirements regarding packaging and labelling. The supplier should adhere to these requirements.

4.15.6: Delivery

QS 9000 standard insists on 100 per cent on-time delivery from suppliers. The supplier must establish systems that enable delivery reliability to be achieved. Corrective action is required for less than 100 per cent on-time delivery. Advance shipment notification should be provided to the customer. Production planning should be applied to firm customer orders. Small lot production or one-piece flow methods are recommended.

4.16: Control of Quality Records

QS 9000 specifies how long records should be held. Quality performance records such as inspection records, statistical process control charts, etc. should be maintained for one year after the calendar year in which they were created. Technical records such as tooling records, purchase orders, parts approval etc., should be maintained for as long as the part is active plus one extra calendar year. Records of management review meetings and internal audits should be maintained for three years. The durations specified are subject to extra requirements from the customer or government regulations.

4.17: Internal Audits

QS 9000 requires that the working environment be audited because of its impact on product quality.

4.20: Statistical Techniques

4.20.2: Procedures

QS 9000 requires that appropriate employees be familiar with basic statistical concepts.

QS 9000: Section 2

Section 2 addresses three areas, Production Part Approval Process, Continuous Improvement and Manufacturing Capabilities.

1: Production Part Approval Process

This section consists of two subsections as follows:

1.1: General

The supplier is instructed to follow the Production Part Approval Process reference manual. The manual documents the type of information required before a part can be approved as acceptable. All new parts or major changes involving a change in processing, material or location must go through the part approval process before it is considered acceptable.

1.2: Engineering Change Validation

All changes to the product or process must be validated according to the Production Part Approval Process manual.

Quality Systems: ISO 9000 and QS 9000

2: Continuous Improvement

This section stresses the use of continuous improvement techniques throughout the supplier's organisation. There are three subsections as follows:

2.1: General

The supplier has to demonstrate work done on improving quality, delivery speed, cost reduction, etc. The supplier must demonstrate continuous improvement in process capability for variable data and continuous reduction in defects per million for attribute data. The supplier should be able to demonstrate a commitment to continuous improvement throughout the organisation.

2.2: Quality and Productivity Improvements

The supplier has to identify improvement opportunities and use appropriate means for achieving the improvements. Documentary evidence of continuous improvement is required for all functions including finance, human resources and marketing as well as design and manufacturing.

2.3: Techniques for Continuous Improvement

The supplier has to demonstrate a good understanding of continuous improvement techniques such as process capability studies, measurement system analyses, benchmarking etc.

3: Manufacturing Capabilities

This section consists of three subsections. It emphasises the need to ensure capability in all areas of manufacturing.

3.1: Facilities, Equipment, Process Planning and Effectiveness

The standard requires that cross-functional teams are used to evaluate and develop the capabilities within manufacturing from facility layouts, balanced product line, automation, ergonomics, etc. The team should identify opportunities for improvement and ensure continuous improvement.

3.2: Mistake Proofing

Mistake proofing or *Poka Yoke* techniques should be used in planning production processes to prevent the production of non-conforming goods.

3.3: Tool Design and Fabrication

The supplier has full responsibility for the design, fabrication and inspection of all tooling and equipment. The supplier also has a responsibility to ensure that customer-owned tooling and equipment is permanently marked and visible.

3.4: Tooling Management

The supplier has to develop an effective system for the management of tools and equipment. The supplier is also responsible for management of the tools and equipment if they are subcontracted. For subcontracted tools and equipment, the supplier must develop a tracking and follow-up system.

QS 9000 Section 3: Customer Specific Requirements

Fulfilment of the criteria documented in Section 3, "Customer Specific Requirements", can only be audited by the customer in a second party audit. It is not auditable by a third party. Section 3 lists the requirements that are specific to each of the companies originating the standard. It covers the areas where harmonisation was not possible. Each of the companies attempts to achieve competitive advantage through the individual requirements specified in this section. Each customer provides a bibliography of manuals which documents their customer-specific requirements. The supplier is advised to contact their particular company for an interpretation of these documents. The supplier is required by the standard to adhere to their requirements and to have copies of the manuals on hand. The following is a list of the manuals generated by each company.

Chrysler has seven manuals:

1. Design review guidelines
2. Design verification plan & report

Quality Systems: ISO 9000 and QS 9000

3. Reliability functions
4. Reliability testing
5. Test sample planning
6. Priority parts quality review
7. Product assurance guidelines.

The Ford Motor Company has nine reference manuals:

1. Manufacturing standards for heat treating W-HTX-1 and W-HTX-12
2. Packaging guidelines for production parts
3. Failure mode and effects analysis handbook
4. A quality, reliability primer
5. QOS assessment & rating procedure
6. QOS quality is the name of the game!
7. Supplier quality improvement guidelines for prototypes
8. Heat treat system survey guidelines
9. Team orientated problem solving.

General Motors has fifteen reference manuals:

1. C4 technology programme, GM-supplier C4 information
2. Key characteristics designation system
3. Supplier submission of material for process approval
4. Problem reporting and resolution procedure
5. Supplier submission of match check material
6. Component verification & traceability procedure
7. Continuous improvement procedure
8. Evaluation and accreditation of supplier test facilities
9. Early production containment procedure

10. Traceability identifier requirements for selected components on passenger and light truck vehicles, traceability identifier requirement

11. Specification for part and component bar codes ECV/VCVS

12. Procedure for suppliers of material for prototype

13. Packaging and identification requirements for production parts

14. Shipping/parts identification label standard

15. Shipping and delivery performance requirements.

These documents change over time and therefore the supplier is recommended to verify annually that they are using the latest revision of the manuals. For the various truck manufacturers, the supplier must contact their purchasing department to determine any additional requirements they may have.

FROM ISO 9000 TO QS 9000

The company should be registered to the ISO 9000 standard before being audited to the QS 9000 standard. The existing ISO 9000 system can be adapted to cater for the extra requirements of QS 9000.

The QS 9000 document is more specific than ISO 9000. For example, 4.4.2.1 of the ISO 9001 document states that: "The design and verification activities shall be planned and assigned to qualified personnel equipped with adequate resources". QS 9000 documents the meaning of the term "qualified". Design staff should be trained in skills such as geometric dimensioning and tolerancing, value engineering, quality function deployment, etc. Major additions refer to the quality planning process using the Advanced Quality Planning and Control Plan and the development of process capability (Cpk) indices which are in general greater than 1.33. The documentation and quality system associated with the ISO 9000 standard will have to be updated to reflect these additions.

The quality manual will have to be extended to include references to Sections 2 and 3 of the standard. Each of these requirements have to be addressed briefly in the manual. For example, the manual should briefly describe how the company adheres to the production part approval manual, how it continuously endeavours to improve its operations through the use of continuous improvement techniques, how it plans its facilities, tool management systems, etc., in accordance with the requirements of Section 2. Similarly, the manual should refer to the customer-specific criteria that the quality system address according to Section 3 requirements.

QS 9000 typically results in a significant investment in training. The standard lists skills that should be available to design groups, including such skills as finite element analysis, simulation, computer aided design, etc. The standard places strong emphasis on the use of statistical methods such as gauge R & R for evaluating measurement systems and Cpk for determining the capability of a process. In general, a broad range of personnel will need to be trained in these techniques in order to establish their use in practice. Training tends to be the biggest cost associated with implementation of QS 9000.

Similar to the recommendations for ISO 9000 implementation, a cross-functional team should be formed to implement the QS 9000 system. The additional requirements are listed and responsibilities to ensure compliance distributed. The implementation group should keep in close contact with their customer to ensure that their interpretation of Section 3 requirements is satisfactory.

Conclusion

Quality systems such as ISO 9000 and QS 9000 provide management with an opportunity to:

- Document their commitment to quality

- Demonstrate that commitment to external customers and internal employees

- Standardise best practice methods where they exist

- Set up a comprehensive system geared to protecting the customer from receiving poor quality product

- Set up systems geared towards continuous improvement.

Each of these endeavours should strengthen the competitive position of the company. However, care is needed during the implementation stage to ensure that the end result is not the establishment of a paper bureaucracy. The aim of the quality system is to ensure quality at an economic cost. Excessive bureaucracy stifles the synergy of an organisation and adds cost to the end product.

Many customers specify that compliance to one of the standards is a precondition for doing business. The big three automotive manufacturers in America have already set deadlines for compliance. As more companies realise that the quality of goods received depends on the quality system in place in the originating factory, they too will expect their suppliers to conform to recognised quality standards.

Chapter 4

EQUIPMENT ENGINEERING AND MAINTENANCE

INTRODUCTION

The emergence of low-cost computers, inexpensive position control systems, pneumatics and other such technologies has led to automated machinery becoming more affordable and more readily available. As companies invest in these new technologies, their ability to compete depends on their ability to manage their equipment's performance. Poor equipment management leads to a poor return on investment.

The equipment maintenance function is often perceived as a necessary evil within companies. When machines are running poorly, they are open to allegations of poor maintenance practice, being under-staffed or poorly trained. When equipment is running well, they may be singled out as being over-resourced and perhaps, considered not to be working as hard as their production colleagues. Maintenance personnel provide a service to production, who, as the customer, may have a subjective perception of the maintenance department. Sometimes it is only when maintenance personnel turn around poorly performing machines that they have their work and abilities properly recognised.

An effective equipment management system seeks to maximise equipment performance by focusing the entire factory on the care of equipment. The performance of equipment depends on more than the competencies of technicians. It depends on factors such as the level of training of the operator, the quality of materials being processed and the original design of the equipment.

Care for equipment should not be delegated solely to the maintenance department. Too often, the role of maintenance is to fix machines when they have broken down or are not performing to specification. The reasons why they broke or why they are not performing within specification are not given sufficient attention. In this environment, problems tend to repeat. Typically, a technician's time is spent performing three types of tasks: simple repetitive tasks, technical repetitive tasks and improvement tasks. Figure 4.1 show a typical distribution of the tasks as part of a weekly workload.

FIGURE 4.1: TYPICAL DISTRIBUTION OF A TECHNICIAN'S WORKLOAD

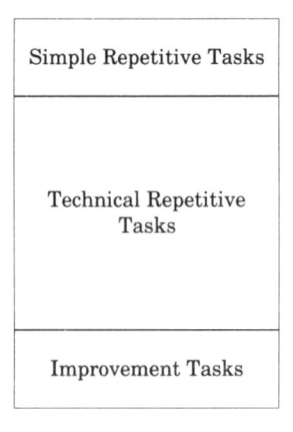

Simple Repetitive Tasks
Technical Repetitive Tasks
Improvement Tasks

Simple repetitive tasks are those which could easily be performed by an operator. For example, a technician in one company always cleans the pipettes on a particular machine before going for lunch. He believes that it reduces the risk of being called to the machine during his lunch break. Cleaning the pipettes is a simple repetitive task.

In another example, a technician performs equipment changeovers — a task that is easy to do, yet time-consuming. The operator could be trained to perform this work. Another technician has to repair the transport system of a machine because the dirt build-up has caused the timing to go out of sync. If the operator maintained a clean machine, this repair would not be necessary.

Equipment Engineering and Maintenance

Technical repetitive tasks deal with problems for which a technically trained person is required. On a seasoned production line, most of the problems can be categorised as technically repetitive. There are few technical faults that have not occurred previously. The fault may be new to the technician attempting to repair the machine, but there may be technicians, perhaps on other shifts, who have previously fixed this problem. Other technical work fitting into this category includes:

- Breakdowns whose solutions are known but difficult to implement (e.g. rebuilding a tooling set up)

- Preventative maintenance and predictive maintenance.

The third form of work is termed improvement work. It includes working on first-time occurrence machine problems, or working on machine improvement projects. When the technician is applying their skills to overcome initial equipment design faults, they are adding value to the company. When they use their technical skills to resolve ergonomic problems, they are improving productivity. The more improvement work performed by the technical staff, the better the competitive position of the company.

Moving the technician's workload towards more improvement work requires the co-operation of all factory personnel. It requires support and direction from top management. Ideally, it means delegating the simple repetitive tasks to production, minimising the time spent on technical repetitive tasks and identifying opportunities for improvement work. Figure 4.2 compares the workload before and after delegation.

There are a number of tools that provide a framework for making this transformation. One of these is Total Productive Maintenance (TPM). Its aim is to focus top management on equipment maintenance and to transfer simple repetitive tasks of production through the use of autonomous maintenance. This is maintenance performed by machine operators. Daily tasks such as machine cleaning, lubricating, checking the condition of various machine components and making simple repairs can be delegated to machine operators.

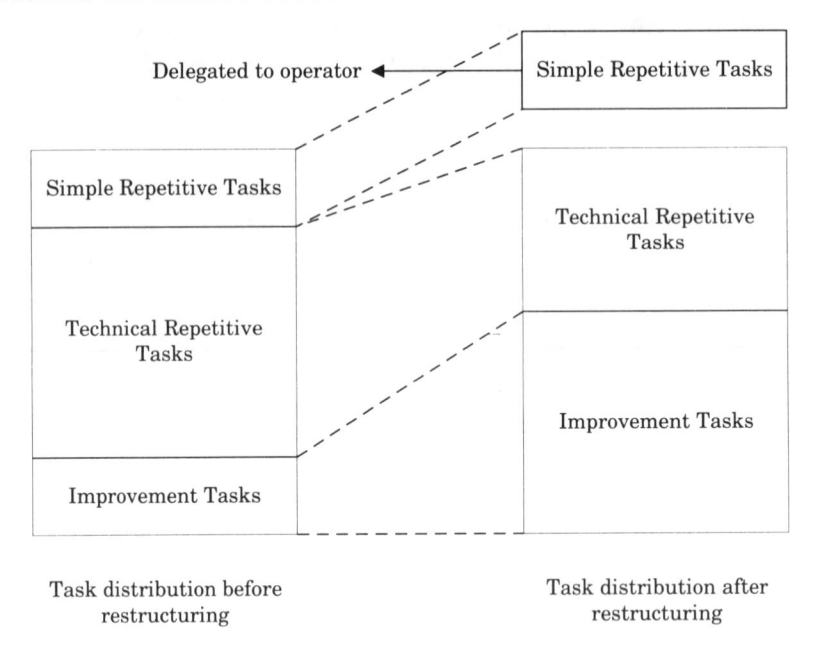

FIGURE 4.2: RESTRUCTURING THE TECHNICIAN'S WORKLOAD

Task distribution before restructuring	Task distribution after restructuring
Simple Repetitive Tasks	Simple Repetitive Tasks (Delegated to operator)
Technical Repetitive Tasks	Technical Repetitive Tasks
Improvement Tasks	Improvement Tasks

TPM also stresses the importance of preventative and predictive maintenance techniques. Much information gathered during autonomous maintenance is translated into preventative maintenance procedures. As more is learned about the causes of faults, these preventative maintenance procedures are updated. Predictive maintenance techniques involve monitoring the condition of equipment in order to predict the onset of failure. Establishing an effective predictive maintenance procedure can save money in part replacement costs and in avoiding downtime.

Reliability centred maintenance (RCM) is a relatively new tool which originated within the airline industry. It involves systematically analysing each component of the machine and determining the best way of avoiding breakdown. Reliability centred maintenance is a very proactive means of establishing an effective preventative maintenance programme.

Techniques such as TPM and RCM reduce the levels of simple repetitive tasks and technical repetitive tasks. The extra time available to technicians should be applied to improvement work. Improvement work can be directed to eliminate some of the causes of breakdown through machine redesign. The ideas of Fitt (1951) and Jordan (1963), who describe how technology can improve productivity, are also presented. This should help to identify areas within production where technology can achieve competitive advantage.

The maintenance department can be perceived merely as equipment repair personnel, or as important strategic contributors to the company's competitive position. As strategic contributors, they perform improvement work; they improve the factory's overall performance by designing out repetitive problems, by making ergonomic improvements, by developing jigs and fixtures and by providing a pool of technical expertise.

DEFINITION OF EFFECTIVE EQUIPMENT MAINTENANCE

Effective equipment maintenance aims to keep equipment in as-good-as-new or even better-than-new condition. It achieves this goal with the minimum expenditure on resources.

One way of achieving this is to sub-contract equipment maintenance to vendors. A service engineer visits the plant at regular intervals, on behalf of the vendors, to perform preventative tasks and upgrade the equipment as necessary. This approach tends to be expensive when the factory has a diverse range of equipment. It also leaves the company exposed to high levels of equipment downtime between service visits.

A more common approach is to train factory technicians to maintain their own equipment. Technicians receive training from the supplier of the equipment, enabling them to fault-find on the machine and perform regular preventative maintenance work.

A comprehensive equipment maintenance policy takes a cross-functional approach to equipment. Equipment is one of several resources that combine to add value to raw materials. Equipment has to be suitable to the capabilities of the operators who work the machine, to the raw materials that it processes and to the environment in which it operates. The contribution that machines make to the overall factory's performance is maximised when the people who are responsible for each function work together to create compatibility between the different factors. Total Productive Maintenance provides a framework which enables the respective functions to more effectively support equipment performance.

INTRODUCTION TO TOTAL PRODUCTIVE MAINTENANCE

Total Productive Maintenance was developed in Japan in the 1970s. It considers equipment maintenance to be the responsibility of the whole factory. TPM focuses factory resources from production, process engineering and maintenance on the objective of achieving zero breakdowns. It uses the metric, overall equipment effectiveness (OEE), to measure equipment performance level. The machine's overall equipment effectiveness is defined as:

OEE = Uptime x Performance x Quality Level

Each of these elements is explained below:

Equipment *uptime* is one of the most common metrics for evaluating machine performance. Uptime (U) is generally seen as the amount of time that the machine is operational. It is calculated as:

$$U = \frac{\text{Available Time} - \text{Machine Downtime}}{\text{Available Time}}$$

(Available time is the amount of time the machine is expected to be utilised).

Small stoppages in the region of 0 to 10 minutes tend not to be considered as downtime by operators, as it is often impractical to individually log each of these stoppages. The uptime metric is easy to understand and trace within the production environment and it provides a lot of information on the level of severe breakdowns occurring.

Another metric used by companies is *operating performance efficiency*. Operating performance compares the throughput from the machine during its uptime to the theoretical output according to its specification run rate. Machines are purchased to produce at a specified run rate, but minor stoppages which are not captured in the uptime metric mean that the machine frequently does not achieve its specified output during operation. Operating performance efficiency (P) is defined as:

$$P = \frac{\text{Actual Units/Operating Hours}}{\text{Theoretical Units/Hr}}$$

(Operating Hours = Available Time − Machine Downtime).

A third metric used by companies is the quality of product manufactured in the machine. Neither of the above metrics take account of the scrap levels generated by the equipment. Scrap production wastes both capacity and money. The quality of output (Quality Performance) is measured as:

$$Q = \frac{\text{Quantity Units Out}}{\text{Quantity Units In}}$$

Each of these metrics describes an aspect of the machine's performance. By multiplying the three metrics together, the overall effectiveness of the equipment (OEE) can be measured as:

$$OEE = U \times P \times Q$$

Maximising the performance of equipment is synonymous with increasing equipment effectiveness. Consider the example of a printing machine in a factory with the following characteristics:

Uptime	0.75
Operating Performance	0.85
Quality	0.90
Equipment Effectiveness	0.57

The equipment is 57 per cent effective; this means that there is ideally scope for a potential 43 per cent extra equipment capacity. Typical values of OEE for companies is in the region of 60 per cent, thereby indicating that there is usually plenty of scope for improving throughput at a minimal extra cost. A value of 85 per cent effectiveness for all equipment is often taken as an initial goal. The equipment effectiveness metric focuses people's attention onto areas of machine performance other than equipment breakdown. Machines that spend their time producing scrap are obviously ineffective. Typically accepted downtime, such as machine set-up time between product runs, will have to be minimised.

Total productive maintenance seeks to maximise the OEE of all factory equipment. Five major reasons are presented for poor performance of machines against OEE:

1. *Equipment breakdowns*: Chronic breakdown due to the failure of machine components.

2. *Set-up and adjustments to equipment*: Time wasted in setting up and adjusting machines between product changeovers. This form of time wasting is analysed in more detail in the chapter on Just-in-Time techniques.

3. *Short stoppages*: Many short stoppages to equipment go unrecorded. Stoppages under ten minutes are often not logged as downtime but they reduce the operating performance efficiency of the machine.

4. *Reduced operating speed*: When machines perform poorly, their speed is frequently reduced in an attempt to minimise the level of defects generated or the occurrence of minor and chronic breakdowns.

5. *Defect units produced on the machine*: Machine time is wasted in processing units that have to be scrapped or reworked.

TPM addresses these problem areas by identifying the causes of problems and establishing preventative measures wherever possible to control the causes of poor equipment performance.

Autonomous maintenance (AM) can be defined as equipment maintenance performed by the machine operator, and is the first step to improving performance. The operator takes responsibility for keeping their equipment and work area clean and tidy. Simple maintenance tasks such as lubrication, checking the condition of parts and performing minor repairs are delegated to the operator. This work is standardised and documented so that there is consistency in how work is undertaken by various operators.

TPM also uses preventative and predictive maintenance techniques to control the causes of poor performance. The causes of variation in performance are analysed and procedures are established to prevent a recurrence.

TPM stresses that top management must actively take part in efforts to improve equipment. They must provide direction and support for the initiative. TPM means that the production team take on some tasks associated with the maintenance department. Production will have to allow this redistribution of responsibilities, and also permit machines to be allocated to maintenance for detailed preventative maintenance work. A cross-functional approach is needed; this will only be successful when it is a management priority.

The other techniques discussed — reliability centred maintenance (RCM), troubleshooting techniques, data gathering and analysis forms — are not generally associated with TPM but they support the goal of maximising equipment overall effectiveness.

SUITABILITY OF EQUIPMENT MANAGEMENT PROGRAMMES

Maintenance techniques such as TPM are applicable to companies with a significant investment in automated equipment. The concepts are relatively inexpensive to implement but require a high level of focus on the implementation area from production, engineering and maintenance personnel. However, the successful conclusion of the project results in increased productivity in both production and maintenance functions. The maximum gain is made when the techniques are applied to a bottleneck operation since any improvement in performance at this function results in increased capacity throughout the line.

As TPM means transferring tasks to machine operators, this will be more easily accomplished where good relations exist between management and operators. Maintenance technicians may be worried about becoming redundant while operators may have concerns about their new role. Maintenance technicians may be sceptical about systematically attempting to eliminate machine problems. A pilot introduction of the new system may help alleviate many of these concerns.

The documented system supports training of new operators and technicians. It helps to make the work environment more consistent and predictable, which aids the introduction of other techniques such as Just-in-Time manufacturing methods.

IMPLEMENTATION GUIDELINES

A model for the implementation of an effective equipment management system is presented below. The model provides a framework for the introduction and description of various maintenance tools and indicates how the tools support each other. However, it is not proposed as a universally applicable model, but rather one that can be adapted to the individual needs of particular industries.

The first stage, *preliminary work*, consists of establishing the scope for improvement in equipment performance. This may involve calculating the overall equipment effectiveness on key equipment or estimating the cost incurred through poor maintenance practices.

The need for top management support is reiterated and the concepts of autonomous maintenance introduced. Autonomous maintenance is about basic care for the equipment by the operator who is using it. It should be considered as part of any maintenance strategy, because keeping equipment clean and lubricated daily is a minimum requirement for proper machine care.

The second stage, *analysis work*, describes:

- A guide to troubleshooting equipment
- Effective data collection and data analysis
- Preventative maintenance techniques
- Predictive maintenance techniques
- Reliability centred maintenance
- An effective spares policy

- Opportunities for productivity improvements through technology.

During the third stage, *implementation recommendations*, suggestions are made which should assist the improvement team.

PRELIMINARY WORK

The first step towards a more effective equipment management system is to establish how well machines are currently performing and the costs associated with this performance. The OEE can be calculated for various pieces of equipment and translated into the cost of lost productivity. The cost of spares and the indirect labour costs within the maintenance department are also related to the effectiveness of the current maintenance system. These costs can be benchmarked against similar industries. A favourable result may indicate that the current system is performing adequately. An unfavourable result may be the stimulus that managers need to concentrate their efforts on a cross-functional approach to maintenance.

The OEE levels define the scope and potential for improvement. Japanese companies adopting TPM consider that an OEE of 85 per cent should be achievable on all processes. The difference between the machine's OEE and 85 per cent is the level of improvement that should be attainable.

A pilot introduction of the new system will help generate acceptance amongst maintenance and production personnel. The machine chosen for the pilot should have a sufficiently low OEE to enable real improvement to be perceived by the implementation team. It will be easier to recognise the merit of the system when the pilot transforms the machine from a poorly operating one to a consistently good one. When choosing the pilot machine, the team should consider the attitudes of the people who currently operate and maintain the machine. Operators with an open-minded, positive attitude will make implementation easier.

The first step recommended by TPM for improving OEE is autonomous maintenance. Autonomous maintenance tasks are performed by the machine production personnel. The maintenance tasks tend to be minor, such as cleaning the machine, lubricating, checking for wear and replacing worn parts. Many of these tasks are basic and could be considered as an essential part of daily machine care.

Autonomous Maintenance (AM)

Autonomous maintenance teaches operators about the machines they work on. AM aims to train production operators to perform daily tasks to help minimise downtime and prolong the life of the equipment. They are provided with training usually associated with the maintenance department — simple things like locating lubrication nozzles, identifying filters, explaining their purpose, and explaining how they affect the machine's performance and how they should be maintained. This preliminary information can be expanded, depending on the capabilities and willingness of the operators. They may be trained to replace various components of the machine, or to detect abnormal operation of the equipment.

AM may raise concerns amongst production staff. Supervisors may be concerned that, if operators are performing tasks typically associated with the maintenance department, productivity levels will deteriorate. However, once established, AM tasks should generally be performed in less than 2 per cent of the time available to operators. An investment of less than 2 per cent in cleaning and lubricating will yield much more than 2 per cent improvement in equipment performance!

Equipment operators will also be concerned about being asked to take on extra responsibilities. They may be worried about their ability to perform the new tasks. The level of anxiety will depend on each factory's circumstances and work environment. However, as automation plays a more important role in the success of factories, it becomes ever more important to share responsibility for equipment upkeep with equipment operators. Just as you would not drive your car without regular servicing, without regularly checking the oil levels, adding anti-freeze to the radiator, adding water to the window washer, cleaning the car, so also, equipment should not be operated without basic machine care procedures.

During autonomous maintenance, the operator learns how to:

- Keep the equipment clean
- Lubricate the equipment, change filters, replace simple worn parts
- Recognise abnormal equipment performance, and take corrective action for easy fixes
- Identify improvement opportunities such as making parts of the machine more easily accessible
- Learn how the various functions of the machine affect its performance and how adjustments should be made.

Autonomous maintenance is a cross-functional team exercise. It begins by restoring the equipment and its work environment to its original condition. This requires the support of production, maintenance engineering, industrial engineering and process engineering. Each group has a role in returning the equipment and the process to its optimum condition. Quality and process engineering ensure that only good units arrive at the workstation for processing. The industrial engineering department works with production to ensure the workstation has ample storage room and is ergonomically designed, and the maintenance and production staff work together to overhaul the equipment.

There are seven steps to implementing autonomous maintenance:

1. Initial cleaning and inspection
2. Elimination of sources of contamination and reorganisation of inaccessible sections of the machine
3. Creation and maintenance of lubrication standards
4. General inspection
5. Autonomous inspection
6. Workplace organisation and housekeeping
7. Fully implemented autonomous maintenance programme.

Although all seven steps are recommended, only steps 1–5 are described in detail here. Step 6 involves using the "five S" system for workplace organisation which is described in Chapter 6 on JIT. Step 7 involves integrating TPM throughout the factory. This is achieved by expanding tasks 1 to 5 in all factory operations and regularly auditing to ensure compliance.

Step 1: Initial Cleaning and Inspection

Autonomous maintenance is kicked off with a thorough cleaning and overhaul session on the equipment. Time is allocated to the TPM team to restore the equipment to its original condition. There are three immediate goals to the cleaning and overhaul exercise:

- To establish a clean machine.
- To list all aspects of the machines that need to be cleaned, screws to be ordered, filters to be replaced, tooling sections that had to be realigned, etc. List the time taken to perform each task.
- To repair equipment when possible and to list all remaining problems.

It may be worthwhile asking the equipment vendor to take part in the exercise. Their service engineer will have much to contribute to the restoration effort and will be able to help the team develop further AM activities.

The general area around the machine should also be cleaned as part of this exercise. Any tools, paper, etc., that are not regularly used should be disposed of or stored elsewhere.

After the exercise, the machine should be run for approximately 1–2 hours and its operation verified. Any problems during this short run should be noted. It may be worthwhile designating some staff to take notes and others to clean and overhaul the equipment.

Step 2: Sources of Contamination and Inaccessible Areas

The cleaning process is made easier when the sources of dirt, such as leaking pipes, are eliminated. Inaccessible areas of the machine also make the cleaning and inspection process difficult.

Lubrication points may need to be relocated to make them more accessible. Installing perspex windows in doors means that covers do not have to be removed to inspect the machine. Very often, broken wiring can be avoided by reorganising the wiring system. The sources of dirt and inaccessible areas are listed and, over time, the improvement team resolves these problems.

Step 3: Creation and Maintenance of Cleaning and Lubrication Standards

The information gathered provides a lot of valuable information regarding the maintenance of the machine. It can be assumed that sections of the machine which were dirty will become dirty again; that any component which was replaced will have to be replaced again at some interval; or that any screws which were loose will become loose again. These assumptions will, in general, be true unless progressive steps are taken.

For each task undertaken during the restoration session, identify the cause of the problem. If the cause cannot be eliminated, then estimate how long it will take before the condition deteriorates. The solution is to repeat the restorative task before the deterioration occurs or remove the cause of deterioration. In this way, if a drive belt is frayed, the lifespan of the new belt should be estimated and its replacement planned before the end of its life. The preventative work is proceduralised and time allocated daily for implementation. It then becomes the responsibility of the operator to ensure that their equipment is clean and correctly lubricated.

Step 4: General Inspection

As operators receive further technical training, they are able to assume more responsibility. Step 4 involves training operators in areas such as pneumatics, hydraulics, transportation systems and minor repairs so that they are capable of inspecting the various components that comprise the machine. Operators are trained on how various machine components — for example, the water trap in pneumatics or the vibration system — affect the machine's performance. Operators should then be able to inspect the correct functioning of these components and make adjustments as necessary.

The lessons learned in the classroom are put into practice at the workstation. The operators inspect the various components of their machine and list any problems discovered. The findings of the inspection process are documented. Areas that should be regularly inspected are listed along with their acceptance criteria. The improvement team analyses the downtime records and defines component checks that may prevent breakdowns.

Step 5: Autonomous Maintenance

Autonomous maintenance involves combining the findings from steps 3 and 4 into comprehensive preventative maintenance procedures. The tasks are divided between operators and maintenance staff, depending on the complexity involved in performing them. Figure 4.3 shows a format for documenting the preventative tasks.

As further breakdowns occur, the team should work together to develop preventative measures to avoid recurrence. The trend in OEE should indicate the success of the group in meeting its goals.

Autonomous maintenance is a powerful tool for making production staff more aware of proper machine care. The cleaning and inspection process teaches operators how to maintain their machine's condition properly. The training provided by maintenance empowers operators to ensure their equipment operates at its top performance. AM reduces the pressure on the maintenance department by delegating traditional maintenance tasks to other, more appropriate departments.

FIGURE 4.3: AUTONOMOUS MAINTENANCE TASKS DOCUMENTED

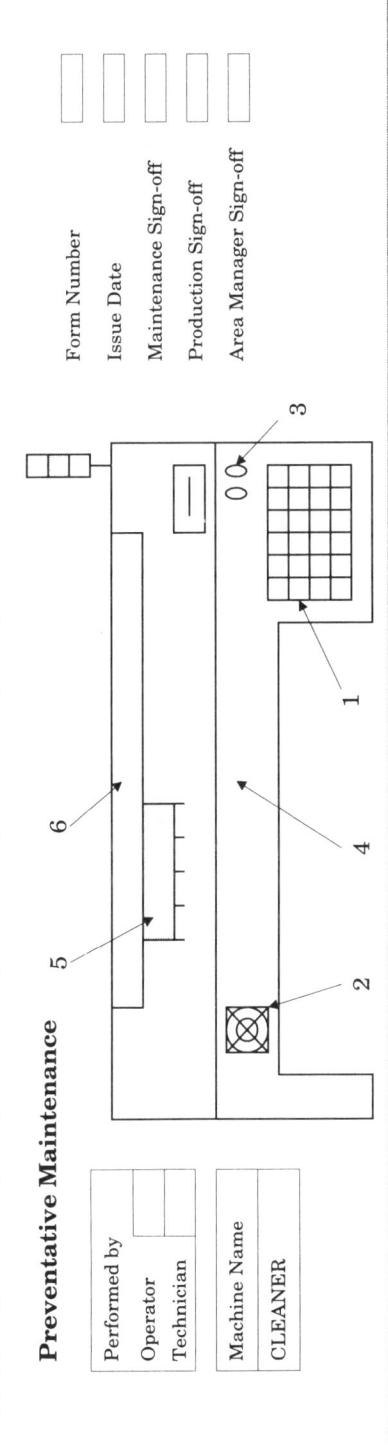

Preventative Maintenance

Performed by
- Operator
- Technician

Machine Name: CLEANER

Form Number
Issue Date
Maintenance Sign-off
Production Sign-off
Area Manager Sign-off

No.	Precautions Power	Precautions Air	Location	Task	Standard	Tools	Response	Time	Freq.
1	Off	—	Air filter	Clean air filter	Dust free filter	Filter, screwdriver	Change filter	5 mins	1/wk
2	—	—	Cooling fan	Inspect for rotation, clean	Rotating smoothly	Visual	Call technician	0.5 mins	1/wk
3	—	Off	Air lubricator	Inspect oil level	Oil above red line	Pneumatic oil	Add oil	1 min	1/wk
4	—	Off	Air regulator	Check pressure level	PSI between red lines	Manual	Move within lines	0.5 mins	1/wk
5	—	—	Machine table	Clean area	Dirt & product free	Manual	Remove stray product/dirt	1 min	1/day
6	—	—	Spray head	Inspect & clean	Nozzles undamaged Nozzles free inside	Wire probe Air gun	Replace damaged nozzles Blow inside nozzles w/air	5 mins	1/wk
7	—	Off	Slide	Inspect and lubricate	Smooth action	WD40 lubricant		1 min	1/wk

ANALYSIS WORK: A GUIDE TO TROUBLESHOOTING

An effective system of fault-finding or troubleshooting should reduce the time needed to resolve technical problems. Fault-finding on equipment is often considered a black art, where some people are gifted while others languish. It is true that some people are very talented at understanding equipment and being able to identify reasons for malfunctions. However, much successful troubleshooting can be achieved by adopting a systematic approach to the study of the problem. R. Pirsig captures the essence of the systematic approach in his 1976 book *Zen and the Art of Motorcycle Maintenance*. The following is an extract from the book and provides ample advice for troubleshooting. Pirsig defines three approaches to equipment repair: inductive and deductive inferences and a formal scientific method.

> Inductive inferences start with observations of the machine and arrive at general conclusions. For example, if the [motor] cycle goes over a bump and the engine misfires and then goes over another bump and the engine misfires and then goes over another bump and the engine misfires and then goes over a long smooth stretch of road and there is no misfiring and then goes over a fourth bump and the engine misfires, one can logically conclude that the misfiring is caused by the bumps. That is inductive reasoning from particular experiences to general truths.
>
> Deductive inferences do the reverse. They start with general knowledge and predict a specific observation. For example . . . the mechanic knows the horn is powered exclusively by electricity from the battery, then he can logically infer that if the battery is dead, the horn will not work. That is deduction.
>
> The formal scientific method consists of breaking the analysis into the following six categories:
>
> 1. Statement of the problem;
> 2. Hypotheses as to the cause of the problem;
> 3. Experiments designed to test each hypotheses;

Equipment Engineering and Maintenance

4. Predicted results of the experiments;
5. Observed results of the experiments;
6. Conclusions from the results of the experiments.

This process is repeated until the problem is resolved.

These techniques help to remove some of the ambiguity involved in equipment repair. The motorcycle example described bumps, while a manufacturing example may consider, say, different batches of material, or different machine operators. The deductive inference necessitates a knowledge of how the machine functions. The scientific method requires imagination, and an ability to test ideas. Together the three approaches provide a useful guide to people involved in equipment repair.

The techniques proved useful in a situation where I was working on a machine problem with some technicians. The machine was not performing its function at random intervals. It had been happening since the second shift on the previous day. A variety of components had been changed but to no avail. The drive mechanism was chain-driven through a series of gear boxes. The drive turned a drum that transferred units during processing. No pattern had been detected to what appeared to be a randomly occurring malfunction. In order to verify the random nature of the problem, we turned the drive shaft by hand, to see if the problem occurred at any regular interval. We wanted to know if there was a pattern, like the bump that Pirsig described in his "inductive inference". We had to manually turn the shaft 176 times between the occurrence and recurrence of the problem. It was then we knew the fault: there was a 1:50 ratio gear box driving a chain. The 176 turns reflected the reccurrence of the problem at approximately the same spot on this chain. The chain had some tight spots due to elongation. It was replaced and the machine returned to production.

Effective Data Collection and Analysis

Replacing the chain in the previous example is only part of the solution process. Downtime caused by problems such as the elongated chain needs to be logged, analysed and controlled.

Unless downtime is properly recorded, and preventative measured taken, technicians will have to repeatedly solve the same technical problems. The minimum information needed to aid permanent resolution of problems is:

- Problem symptom
- Problem cause
- Problem solution
- Downtime.

Figure 4.4 shows a suitable format for collecting this information.

In order to minimise the time spent performing repetitive technical work such as troubleshooting reccurring problems, the information should be converted into an easy-to-use troubleshooting guide. A troubleshooting guide lists the machine problems, their possible causes and solutions. It acts as a quick reference for technicians and can be an invaluable aid when there are a multitude of possible causes to particular problems. It enables the technician to systematically investigate each previous cause before exploring for new causes. This tool effectively reduces the time spent resolving technically repetitive problems.

Whenever possible, the information in the downtime records should be translated into preventative measures that can be taken to avoid reccurrence of the problem. In the case of the elongated belt, preventative work may include replacing the belt or checking it for tight spots at regular intervals. These preventative steps minimise exposure to problems.

Preventative Maintenance

By changing the chain at intervals recommended by the belt manufacturer or according to historical experiences, exposure to a recurrence of the problem is minimised, if not eliminated. The life of the chain would be increased if it were properly

Figure 4.4: Equipment Data Collection Chart

Equipment Management Form   Machine:

Date	Down @ time	Operator	Problem Description	Cause of downtime	Solution	Up @ time	Technician

Figure 4.5: Preventative Maintenance Calendar

Machine

Task	Week							
	1	2	3	4	5	6	7	8
	Tech. Initials							
Lubricate Bearing								
Grease Slides								
Clean Vacuum Nozzles								
Clean Filters								
Inspect/change gear box oil								
Replace drive chains								
Etc.								

lubricated at regular intervals. Autonomous maintenance stresses the importance of analysing downtime logs continuously and updating preventative maintenance procedures when required.

Most equipment vendors recommend a list of preventative tasks to be carried out on the equipment at regular intervals. Their tasks should also be included as part of the preventative maintenance procedures.

Predictive Maintenance

Predictive maintenance tasks aim to predict the occurrence of problems so that action can be taken to avert downtime. The most common usage of predictive maintenance is with wear parts. Variation in the key dimensions of wear parts are measured and, when they go below a predefined level, the part is replaced. In this way, downtime and poor machine performance are averted.

There are two stages to predictive maintenance: first, identify a characteristic that varies as a component nears its end-of-life; and second, define the level at which the component should be repaired or replaced. For example, when a bulb is nearing its end-of-life it may gradually reduce its operating temperature before complete failure. By understanding the failure mechanism and the characteristic associated with its failure, the breakdown can be predicted. The temperature of the bulb is measured from a certain distance at regular intervals and logged, and temperature variations associated with failure are noted. A procedure is set up such that any recurrence of these temperature variations with future bulbs results in the earliest possible replacement of the bulb.

In the example of the chain, the elongation could be determined by measuring displacement of different parts of the chain from a fixed point. The elongation could also be determined subjectively by feeling for variations in tightness or slack. The following is a list of characteristics and the techniques for measuring them. If a predefined level of the characteristic is linked to component failure, then the failure can be predicted by monitoring the characteristic. A sample of characteristics and their measurement instruments are listed below:

Characteristic	Measurement instrument
Speed variation	Tachometer
Torque	Torque meter
Pressure	Pressure gauge
Noise	Sound level meter
Vibration	Vibration amplitude/frequency analysers
Dimension	Micrometer, shadowgraph

Reliability Centred Maintenance (RCM)

Reliability Centred Maintenance is a highly proactive approach to developing a comprehensive preventative maintenance policy. RCM was developed in the commercial airline industry as a means of reducing the time spent on servicing airplanes while at the same time improving the level of service performed. The servicing of the airplanes was based on the concept of the reliability bath tub curve. The bath tub curve suggests that parts will have a high failure rate during early-life and end-of-life phases (see Figure 4.6).

FIGURE 4.6: BATH TUB RELIABILITY CURVE

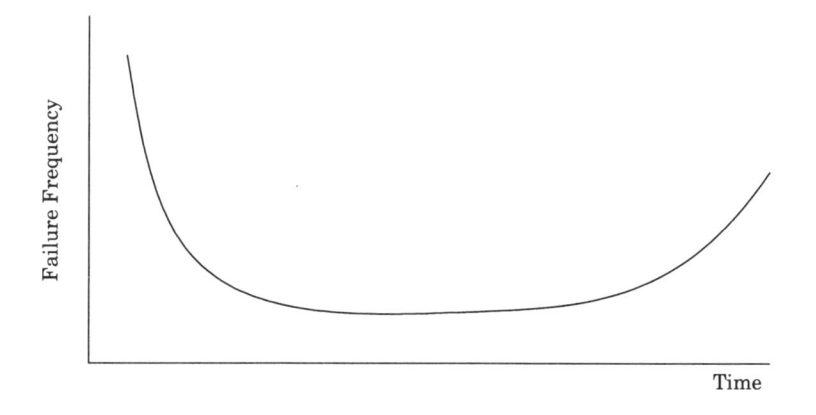

During the servicing of the airplanes, many parts were being replaced in anticipation of their reaching end-of-life. A study performed by the airline industry found that 89 per cent of these parts did not follow the typical bath tub curve; they showed different patterns of failure. Figure 4.7 illustrates some of the reliability curves associated with machine components.

FIGURE 4.7: OTHER RELIABILITY CURVES ASSOCIATED WITH MACHINE COMPONENTS

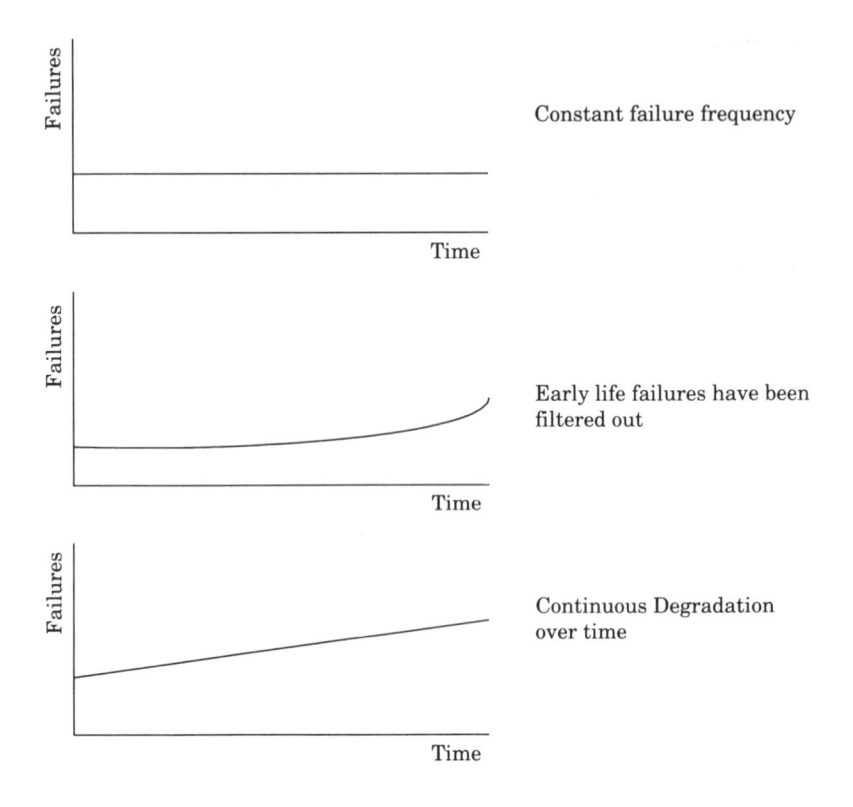

Constant failure frequency

Early life failures have been filtered out

Continuous Degradation over time

These are some of the reliability curves which describe component behaviour. They differ from the traditional bath-tub curve, so the concept of regular replacement of parts in order to minimise breakdowns would be ineffective. A new approach was developed called Reliability Centred Maintenance which uses a structured approach for determining the preventative maintenance requirements of equipment. It seeks to maintain the "inherent reliability" of equipment. A group of people who are closely associated with the equipment, such as the team that established the autonomous maintenance activities, analyses the machine under the following headings:

List the significant functions that constitute the machine:

- The machine transportation system
- The labelling section
- Etc.

Define the acceptable performance criteria for these functions:

For the transportation system, the product must move at a linear speed and no jerky motion is acceptable (it is currently not possible to quantify acceptable tolerances).

List the consequence of the function failure:

Are there safety or economic implications, or hidden elements where the consequence of failure may not be immediately detected? For example, failure of the transportation system results in scrap product, resulting in an economic loss.

List the causes of each functional failure:

Elongation of any of the drive belts, wear out within the gearboxes, etc.

Identify what can be done to prevent failure:

For drive chains:

- Weekly lubrication extends their life;
- Weekly manual inspection of chains for tight spots;
- Replace drive chains every six month.

For gear boxes:

- Inspect the quality of lubrication oil every month;
- Inspect gear box for smooth action;
- Change the oil at least every three months;
- Replace the gearbox every two years;
- Etc.

Depending on the answers to the analysis, at least one of the following four options is taken:

- On-condition task maintenance;
- Scheduled restoration;
- Scheduled discard;
- Default task.

On-condition maintenance means inspecting the equipment components at regular intervals to find and correct potential faults. In the example, on-condition tasks include inspecting the drive chains and the gear box.

Scheduled restoration involves reworking the equipment components at specified intervals to reduce the potential for failure. The example refers to applying lubrication to the chains and changing the oil in the gearbox.

Scheduled discard involves replacing the components at regular intervals in anticipation of failure. The example requires that the chains be replaced every six months and the gear box every two years.

Default tasks involve either redesigning the machine function if none of the previous tasks can suitably control exposure to failure, or alternatively not performing any scheduled maintenance where the cost of preventative maintenance is higher than the cost of rectifying the failure and its consequences. The example described above does not have any default tasks associated with it.

The team considers the failure mechanism and the associated reliability data, and decides on appropriate preventative measures. Historical records of machine performance, advice from vendors and generally the support of an RCM specialist who is familiar with the typical reliability characteristics of equipment components, are recommended to determine the appropriateness of each preventative maintenance task.

RCM is a proactive approach to designing a preventative maintenance system. Each failure mode and its causes are analysed and the most suitable preventative tasks undertaken. Initially, it means intensive analysis of equipment to establish the tasks to be performed at intervals. It tends to result in a

Equipment Engineering and Maintenance 155

well-focused preventative maintenance plan that minimises breakdowns.

Improvement Work

As the techniques discussed in previous sections begin to improve the OEE levels of equipment, technicians will have more time for improvement activities. Problem-solving techniques such as brainstorming or cause-and-effects analysis can lead to permanent solutions for repetitive problems. Technicians have more time to partake in the design of new equipment and to ensure its ease of maintenance. There are a broad range of areas where technical staff can contribute beyond machine repair, and such expertise should be channelled into making the factory more competitive.

In order to effectively utilise these technical skills, the scope for improvement has to be established. The initial task is to systematically identify tasks that would benefit from the application of technology.

Productivity improvements are achieved in two ways:

- Reduce the operator workload by reducing the average time between machine assist or machine/operator interaction.

- Reduce the operator workload by introducing automation and semi-automation technologies to manual operations.

Technology can be applied to manual operations to reduce the level of work involved. Simple tools, such as pneumatics, jigs, fixtures and other aids can increase productivity. Semi-automation principles can be applied to sub-tasks which constitute part of the overall operation. For example, Figure 4.8 represents a situation in one factory which shows how an operator has to switch a valve by hand while manipulating the tool during a repetitive manual operation. The action to switch the valve is moved to a foot pedal. The old method had poor ergonomical features. The new method is easier to use and increased the operator's productivity by 15 per cent! Significant productivity gains can be achieved from a comprehensive study of all tasks performed by operators.

FIGURE 4.8: SWITCHING VALVE

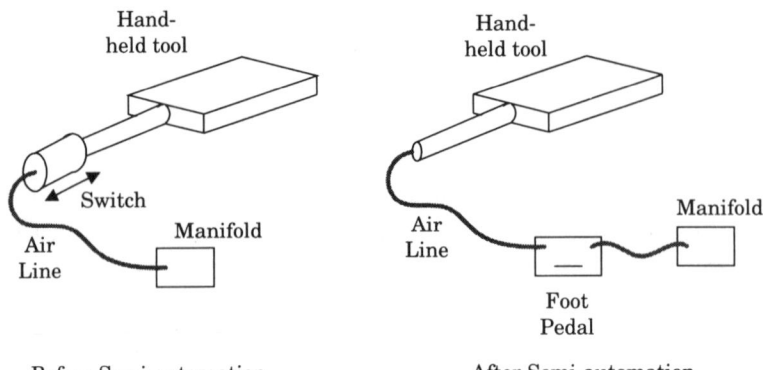

Before Semi-automation — After Semi-automation

Productivity gains are made when the operator's workload is reduced. Jordan (1963) claimed that the basic advantages of the operator over automated procedures were flexibility and adaptability. These are typically non-quantifiable attributes. Jordan considered that if the task can be stated exactly then a machine will do a better job because it is generally more consistent and reliable. Using Jordan's guidelines, all repetitive stagnant operator tasks are candidates for automation. Pure automation can be expensive, but semi-automation, where technology helps the operator perform the task, can be cheap and very rewarding.

In 1951, Fitt listed the relative advantages between people and machines. This list is still relevant today, even though great strides have been made in technology since then. Table 4.1 describes the merits of people and machines under the different properties.

An industrial engineering study can ascertain the distribution of tasks on a production line. Tasks such as loading and unloading a machine can very often be described as repetitive and quantifiable and are therefore suited to automation. Tasks like these are candidates for automation or semi-automation.

TABLE 4.1: FITT'S LIST REGARDING THE SUITABILITY OF AUTOMATION

Property	Machines	Human
Speed	Faster	Slower
Power	Consistent at any level.	1.5 kW for about 10 sec
	Large constant standard forces	1/3 kW for continuous work
Consistency	Ideal for routine, repetition, precision	Not reliable, should be monitored by machine
Complex activities	Copes with tasks in parallel	Copes with one task at a time
Memory	Best for literal reproduction and short term storage	Large store, multiple access; better with principles & strategies
Reasoning	Good deduction	Good induction
	Tedious to reprogramme	Easy to retrain
Computation	Fast and accurate	Slow, subject to error
	Poor at error correction	Good at error correction
Input detection	Some outside human senses, e.g. detection of radio waves	Wide range of stimuli handled
	Insensitive to extraneous stimuli	Sensitive to cold, heat etc.
	Poor pattern detection	Good pattern detection
Overload reliability	Sudden breakdown	Progressive deterioration
Intelligence	None	Can cope with the unpredictable, can anticipate
Manipulation	Specific	Great versatility and mobility

FIGURE 4.9: AUTOMATING MACHINE LOADING AND UNLOADING

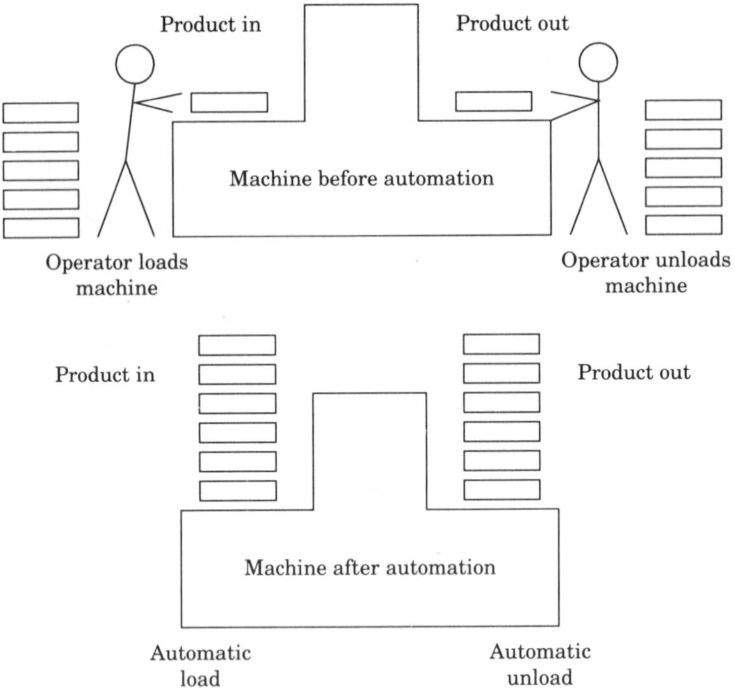

The scope for productivity improvements through the application of technology is readily identified in an industrial engineering study where operations and sub-tasks are analysed under the criteria established by Jordan and Fitt.

IMPLEMENTATION CONSIDERATIONS

Perseverance and dedication are the two essential components of success in equipment management. It takes a lot of effort and discipline to properly maintain equipment. It can be difficult to introduce these new concepts into a work environment where equipment is performing poorly. Technicians and operators are under pressure to keep machines operational. The new techniques are extra work, but successful implementation brings consistency and predictability to the work environment.

Measuring the OEE is the recommended first step. OEE is equal to the product of equipment availability by performance efficiency by quality level (OEE = $U \times P \times Q$). The equipment availability is easy to determine by recording the downtime. When units are lost in the machines, counters may have to be installed in order to quantify the level of defects produced. Inspection records after the process can also be used to measure the level of rejects generated. The performance efficiency can be determined from the number of units processed.

In a JIT environment, it may be awkward to measure OEE, because the machine may not have an operator dedicated to using it full-time. Under JIT, the machine is only used to produce product just in time for subsequent operations. Only the bottleneck operation can be 100 per cent utilised. Demand for the machine's product may be sporadic, in which case the utilisation of the machine will be sporadic. The downtime records will have to clarify whether the machine is not being used because it is broken or because there is no demand for its services.

There are five principal reasons for poor performance against OEE, as described earlier. The improvement team should use problem-solving techniques such as brainstorming to develop ways of overcoming these problems. For example, machine adjustment as a source of downtime is often easy to improve. Very often, operators tend to "tweak" settings in order to run a particular product or to improve the operation of the machine. A procedure would be implemented whereby experienced technicians and operators must mark each knob so that settings are specified for each possible situation. Inexperienced operators will then be able to adjust the settings within specified limits.

Short stoppages can be due to minor breakdowns or to the loading of processing materials by the operator. This form of downtime often becomes the norm and is not perceived as an area for improvement. On one occasion, we trained some operators to calculate the OEE for their machines. An industrial engineering study had indicated the theoretical cycle time. The results showed that the machine was operating at less than 70 per cent. The operators were amazed because they had hit a very high number for the day, achieving the best performance of the machines. They questioned the validity of the industrial engineer's measurement of the theoretical cycle time. The engineer went through his study with the operators and verified his cycle time measurement. The operator measured the OEE again and this time noticed the negative effect that loading the processing materials onto the feed system and minor stoppages had on the OEE. While operating their machine, they experimented with machine settings. Within one week, they managed to set a new record production level for their machines that was 20 per cent higher than the old record. The OEE increased correspondingly. These minor stoppages in the feed system and in loading processing materials had been accepted as the norm even though they were easily improved upon.

Autonomous maintenance will require technicians to spend more time training operators on how to care for equipment. For this reason, technicians may need to develop training skills in order to pass on their technical knowledge.

TPM places strong emphasis on preventative maintenance practices. Autonomous maintenance by production staff and technical preventative maintenance personnel are the cornerstones of a successful equipment management strategy. Many machines come with in-built software for helping the preventative process. The software automatically stops the machine after a predetermined amount of production time and lists the preventative tasks that need to be completed become recommencing production. This software also aids in diagnosing the causes of breakdowns.

An effective equipment management system makes efficient use of technician resources. It reduces the level of technician work involved in breakdown repair through the use of preventative and predictive techniques. The temptation may exist to redirect the newly available technical resources away from the line. However, as this could be a short term strategy, a more appropriate approach may be to focus on opportunities for equipment and productivity improvements through the use of technology. Benchmarking maintenance performance against competitors and/or other factories is another means of developing improvement opportunities.

CONCLUSION

Effective management of technical resources becomes ever more important as industry continues to invest in automation. Equipment management policy directly affects the competitive position of these companies. Effective management means that machines operate to their full potential. It also means that technical resources are utilised to their maximum capabilities. The factory ideally operates in an environment where machine improvement rather than breakdown repair constitutes the main area of work.

There are several modern techniques which enable the maintenance group to reduce the level of breakdown repair and concentrate on improvement work. Total productive maintenance and autonomous maintenance enable operators to take more responsibility for the care of equipment. They are trained to perform simple repetitive maintenance tasks. Techniques like preventative and predictive maintenance, and reliability centred maintenance reduce the level of equipment breakdowns. Improvement work is often self-evident in the manufacturing environment. However, the ideas of Fitt and Jordan will help in identifying those operator tasks which would benefit from the application of semi-automation techniques.

The transition from constantly performing breakdown repairs to introducing improvements results in the factory making competitive gains that their competitors will find hard to imitate. It leads to more job satisfaction for both the operators and the maintenance staff. It results in a continuous improvement in the performance of equipment and the factory.

Chapter 5

JUST-IN-TIME PRODUCTION

INTRODUCTION

In the late 1970s and early 1980s, industry looked to computer software systems such as MRP II (Material Resource Planning) to control their complex manufacturing processes. These systems could analyse customer orders and develop schedules for raw material deliveries and production lines. However, many of these systems were as complex to manage as the production lines. They were poor at coping with the erratic nature of production and, because of quality variation and equipment downtime, schedules had to be continuously revised.

It was during this time that Western industry came under sustained competitive pressure from Japanese manufacturers who were able to produce quality goods at much lower prices. Japanese companies had not invested in software systems or advanced manufacturing technologies, yet they were producing quality product at low cost with minimum inventory levels within their systems. The Japanese had gained competitive advantage by converting their manufacturing processes into a competitive weapon. While these Asian manufacturers were demonstrating that there does not have to be a trade-off between cost and quality or between low inventory levels and factory responsiveness, Western manufacturers were struggling to survive.

In the 1950s, Japanese industrialists had visited American companies to observe manufacturing processes. Now, the situation was reversed and Western manufacturers were visiting their Japanese counterparts to discover how to solve their complex manufacturing problems. Of the many lessons learned, perhaps the most important was the concept of Just-in-Time manufacturing (JIT).

The visitors observed simplified production systems operating in a controlled, predictable environment. The techniques used to achieve these systems include structured flow manufacturing, the kanban system and JIT deliveries. These techniques combined constitute a JIT system. Implementation of the techniques results in reduced manufacturing lead-time, reduced inventory levels and a simplified production system.

In 1969, Skinner had warned of the importance of manufacturing in an article in the *Harvard Business Review*: "A company's manufacturing function typically is either a competitive weapon or a corporate millstone. It is seldom neutral." For the Japanese, it became their competitive weapon, but for western manufacturers, it was a corporate millstone. The Japanese have succeeded in dominating manufacturing in consumer electronic goods and have gained market share in the automobile industry by having the best manufacturing system.

TABLE 5.1: TYPICAL BENEFITS OF JIT

Reduced Manufacturing Lead Time	50%
Increased Productivity	10%
Improved Quality	20%
Reduced Inventory Levels	50%
Reduced Equipment Set-up Time	70%

DEFINITION OF JUST-IN-TIME PRODUCTION

Toyota, the company that first developed the JIT concept, define it as a system for eliminating waste within production. This means that the minimum time, tasks, labour, finance, and equipment processing time should be spent on processing parts. Toyota uses techniques of structured flow manufacturing, Kanban and JIT deliveries to achieve these goals. These are continuous improvement process techniques which, over time, continue to streamline the operations, making them leaner and more efficient.

Structured flow manufacturing involves reorganising equipment and product scheduling. A technique known as Group Technology is used first, to organise production parts into "families" with similar production requirements; equipment and operator resources are then organised to cater for the production of the separate families of parts. Effectively, mini-production lines are set up to produce them. In general, equipment changeover time from one product to the next is reduced because of similarities between parts. The workload is balanced between the different workstations and operators within the mini-production lines, creating a smooth flow of product from beginning to end.

The Kanban system links operations or production lines so that they work in harmony together at a regular pace. It is a communication link which allows the final assembly line to pull product from its various feeder lines and its own operations. In this way, inventory is only processed on a production line as it is required at the subsequent station.

Just-in-Time delivery means that suppliers deliver quality goods just as they are needed on the production floor. As raw materials are used immediately, there should be no need for incoming warehousing, and any defects present are quickly noticed. A faster feedback loop develops with the supplier and ideally any quality issues are resolved before the next delivery.

JIT: NET INVENTORY TURNS AND REDUCED MANUFACTURING CYCLE TIME

The JIT system simultaneously increases the net inventory turnover per year and reduces the manufacturing cycle time. It achieves this by eliminating delays within the manufacturing system. When time delay is eliminated, inventory utilisation increases, the level of value-added work increases and manufacturing cycle time decreases.

The total manufacturing cycle time to produce the end product can be considered as:

Total Manufacturing Cycle Time
= Raw Material Inventory Storage Time
+ Total Internal Manufacturing Time.

Raw material storage time is simply the average time spent in storage. This is reduced by having more frequent delivery of goods from suppliers.

The total internal manufacturing time can be further divided as follows:

Total Internal Manufacturing Time
= Delays between operations
+ Delays within operations
+ Processing time.

Delay between operations relates to all the time lost while the batch of product is not being processed. It includes the time spent in inventory buffers, or the time spent in being transported from one operation to the next. This is minimised by defining maximum inventory levels permitted between operations and controlling these levels with kanbans. (A kanban is a request card which sends a signal to fulfil an order. The kanban system is dealt with in more detail later in the chapter.) Time lost in transport is reduced by bringing the equipment to produce the part closer with structured flow manufacturing.

Delays within operations are the time delays that occur in a lot of, say, 100 units, where 99 units wait while one unit is being worked on. This delay is eliminated by reducing the lot size to one.

Processing time is the time spent performing the tasks that add value to the raw materials. It defines the shortest possible internal manufacturing lead-time.

FIGURE 5.1: TOTAL MANUFACTURING CYCLE TIME —
RELATIONSHIP BETWEEN PROCESSING TIME AND TIME DELAYS

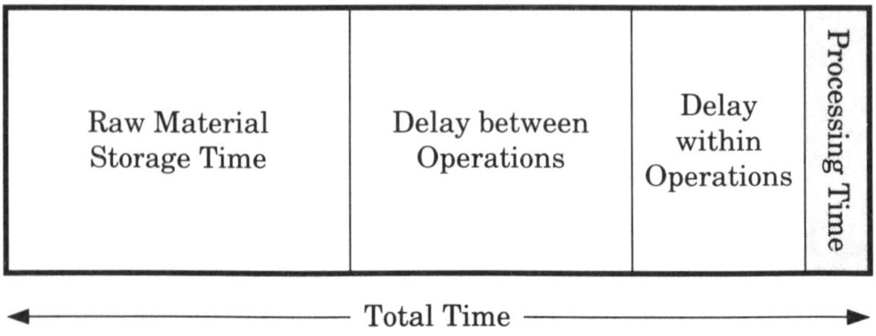

Raw Material Storage Time	Delay between Operations	Delay within Operations	Processing Time

◄──────────── Total Time ────────────►

The total time determines the company's responsiveness and flexibility.

Total time = Raw Material Storage Time
+ Total Internal manfacturing Time

JIT methods control all aspects of the total manufacturing lead-time, including the raw material storage time. It strives to continuously minimise delays within the system. Structured flow manufacturing and the kanban system minimise delays within the internal manufacturing system. JIT deliveries reduce raw material storage time. The net effect is reduced manufacturing lead-time, improved factory responsiveness and increased net inventory turnovers per year.

SUITABILITY OF JIT

JIT is a low-cost improvement programme which improves the overall performance of the factory. It reduces manufacturing lead-times, reduces working capital through increased inventory turnover and it simplifies the production process. It is highly applicable to all medium- and high-volume producers, enabling them to streamline their operations. "Small job" factories producing very low volumes of a very diverse product range will probably derive least benefit from the ideas of JIT.

There are three elements to JIT: Structured Flow Manufacturing; the Kanban system; and JIT deliveries. The three techniques complement each other and, combined, dramatically reduce total manufacturing lead-time. The remainder of this chapter describes each of these techniques in detail.

FIGURE 5.2: COMPONENTS OF JIT

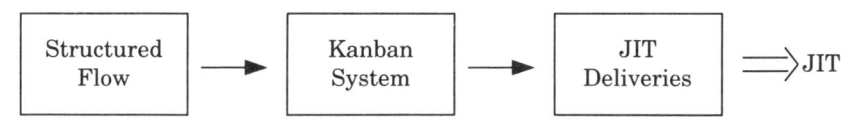

Structured Flow → Kanban System → JIT Deliveries ⟹ JIT

Achieving Structured Flow

A manufacturing system defines how the various input resources are organised to produce the end product. There are three types of manufacturing systems which dominate the organisation of resources in industry: dedicated line production, process-oriented production and structured flow production.

With *dedicated production lines*, machines are laid out on the factory floor in the sequential order required to produce a part. In this system, material moves from the supplier through the production process to the end-user in the shortest time-span and travel distance. Inventory levels within the process are low. The pace of production is dictated by the production rate of the slowest machine in the line and only one range of products are built on each line. This means that dedicated line production requires high volume, repetitive manufacturing to justify the capital investment in equipment.

Management and operators are typically organised into teams with responsibility for the performance and improvement of the production line. Within this structure, responsibility for achieving factory business objectives regarding cost, on-time delivery and quality, for example, are the direct responsibility of the line managers.

FIGURE 5.3: DEDICATED PRODUCTION LINES

PRODUCT LINE 1
Lathe — Mill — Form — Polish — Paint

PRODUCT LINE 2
Form — Weld — Polish — Pack

PRODUCT LINE 3
Drill — Assembly — Paint — Print

PRODUCT LINE 4
Lathe — Mill — Assembly — Polish

Small Receiving Area — Small Shipping Area

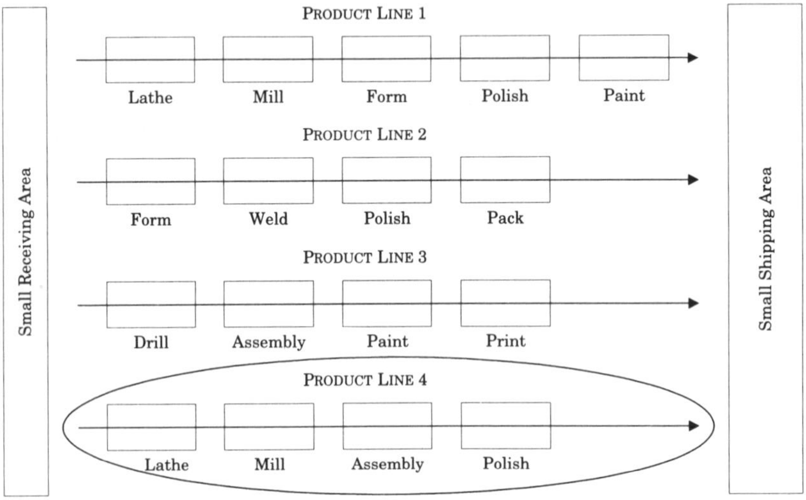

Manufacturing management generally agree that the best performance from a business perspective is achieved when all re-

Just-in-Time Production

sources necessary to supply a product or family of products are grouped together in a close-knit production team. The easiest way to avoid confusion and uncertainty in manufacturing processes is to bring all machines and people involved in producing the complete product into close working proximity. Material and information flow is simplified and employees develop an improved understanding of problems and solutions from working closely to produce the complete product. Dedicated line production is therefore very efficient and effective at producing low-cost, high-volume parts.

Companies producing a wide variety of parts in low and medium volumes tend to use a manufacturing system called *process-oriented production*, which refers to a system where similar machines are located together in one area and batches move between each machine area.

Locating machines according to their function means, for example, that a machining factory positions all lathes in one area, milling machines in another, and grinders, drills, and other equipment in their own area.

FIGURE 5.4: PROCESS-ORIENTED LAYOUT WITH PRODUCT FLOW ROUTES SHOWN

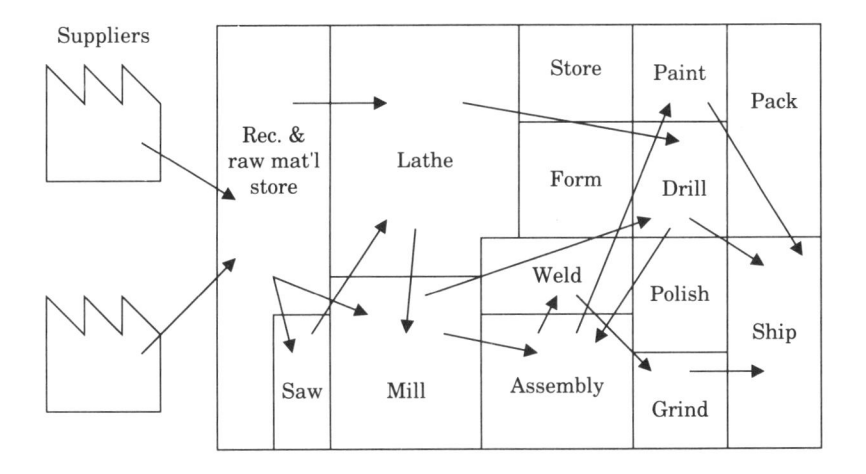

There are certain advantages to this approach. Initial equipment installation costs tend to be lower because the necessary equipment facilities need only be located in specified factory

areas. Also, when a factory is in start-up mode, operators working similar machines learn from each other.

There are also significant disadvantages to this manufacturing system. Transportation distances for parts moving between equipment are much greater. This leads to non-value-added material handling tasks. Product is produced in batches and held in storage until each equipment centre is ready to process it. This results in high inventory levels being stored on the production floor. Personnel resources are linked to equipment centres instead of products and their focus is on the centre's productivity and machine utilisation rather than on ensuring lead-times are reduced or inventory buffers minimised. Their focus is on having a high production figure at the end of the shift. High inventory levels waiting at machines are encouraged, as it prevents disruptions to other operations and sometimes minimises the cost of equipment changeovers.

Structured flow manufacturing is a relatively new system which enables low- and medium-volume producers to adopt many of the characteristics of dedicated line production. It involves categorising the various parts produced within the factory into groups and rearranging equipment and people into mini-production lines dedicated to producing specific families of parts. These mini-production lines provide many of the benefits associated with dedicated line production. There are a variety of tools which facilitate the transition from process orientated manufacturing to structured flow manufacturing. A few of these techniques are shown in Figure 5.5.

FIGURE 5.5: COMPONENTS OF STRUCTURED FLOW MANUFACTURING

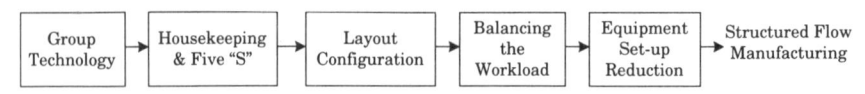

Group Technology → Housekeeping & Five "S" → Layout Configuration → Balancing the Workload → Equipment Set-up Reduction → Structured Flow Manufacturing

The technique known as *group technology*, introduced earlier, is used to break down a range of products into families with similar production requirements. Hundreds or even thousands of parts can be analysed and segregated according to production requirements.

Machines are then divided up and allocated to the production of each group of parts. The machines are laid out in mini-production lines, in the same sequence required to produce the family of parts.

An effective housekeeping policy is a recommended part of any revised layout plan. It helps to ensure the new system is well organised and provides a good environment to work in. The Japanese system called "five S", which is described later, systematically ensures good housekeeping practices.

Group technology helps to reduce time lost in equipment set-up; since many parts within the same family may have similar machine set-ups; also product changeover time is often reduced or even eliminated. A set-up reduction programme can further reduce equipment set-up time. This involves analysing the tasks required to change over equipment and developing creative ways of minimising the time taken.

FIGURE 5.6: PARTS GROUPED INTO FAMILIES

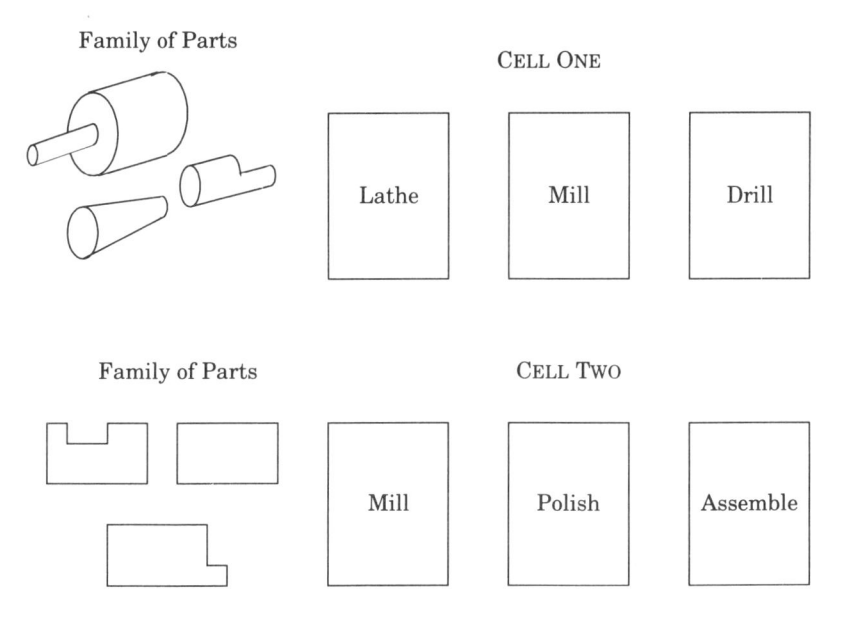

Family of Parts	CELL ONE		
	Lathe	Mill	Drill

Family of Parts	CELL TWO		
	Mill	Polish	Assemble

The manufacturing mini-lines are staffed by multi-skilled operators who have responsibility for the quality of each individual task performed and the overall quality of the finished product. Variations in demand for a family of parts can be accommodated by allocating or removing operators from the mini-line. The production team distributes the work content within the cell to meet the daily production demand.

Constraining equipment within the mini-lines is easily identifiable and the daily production of each line is planned to maximise the performance of these constraints.

Structured flow manufacturing physically brings together a team of people who will work to improve product quality, inventory levels and other typical production issues. It effectively creates mini-factories on the shop floor with the same goals as the overall factory. Structured flow manufacturing improves job satisfaction by giving operators responsibility for producing a complete part instead of merely performing an obscure isolated task. Operators are encouraged to expand their skills and to partake in decision-making within the line. The system results in reduced inventory, reduced manufacturing lead-times and often improved employee morale.

There are other benefits also. The layout sequence of machines reduces travel and lead-times for parts. Large interdepartmental inventories are replaced with small buffer stocks between machines within the mini-line. Scheduling is easier because only the relevant family of parts is scheduled into the mini-line configuration designed for that family. The available capacity is easily evaluated by analysing the capability of the bottleneck equipment in each mini-line. Members of the mini-line become responsible for the quality of the end product, for inventory turns and on-time delivery of customer orders.

The following model is proposed as an implementation guide. It should be a useful guide against which to add or subtract according to the specific requirements of a company.

Preliminary work involves determining the benefits and costs associated with structured flow manufacturing.

Analysis work involves:

- Applying group technology
- Housekeeping and the "five S" system
- Layout configuration

- Balancing the workload
- Equipment set-up reduction

Implementation recommendation and pitfalls describes how various factories carried out the above tasks.

Preliminary Work

Benefits from Structured Flow Manufacturing

There are a host of company objectives which are supported by the concept of structured flow manufacturing. For example, the company may be aiming to reduce manufacturing lead or cycle times. Since structured flow manufacturing brings the machines together to make the part rather than bringing the part to the various machining areas, manufacturing lead-time is reduced. It is possible to gauge the potential for reducing manufacturing lead-time by establishing the ratio of value-added time to total manufacturing lead-time for the factory:

$$\text{Value Added Time Index} = \frac{\text{Total Value Added Processing Time}}{\text{Total Actual Lead-time}}$$

Low values of this ratio indicate that there is plenty of opportunity for improving manufacturing lead-times. With structured flow manufacturing, the reduced lead-times mean that the number of inventory turns per year increases. Investment in working capital is lower and financial ratios associated with the company are improved.

Companies with a diverse product range often experience problems with production scheduling. Structured flow manufacturing means that the products are classified for particular mini-lines, making them easier to schedule. Each mini-line has its own bottleneck; by scheduling the bottleneck, the mini-line tends to be scheduled. The team working on the line is an added resource for schedulers, helping them to ensure on-time delivery of parts.

Structured flow also helps scheduling by reducing time wasted in changing equipment over from one product type to the next. High equipment changeover time often means that the scheduler must decide how much of a particular part they want each month and schedule to produce the month's entire quantity in one production run. By reducing the changeover time for equipment by half, the scheduler can theoretically produce the month's volume in two blocks of production without incurring any extra time lost to changeover. If the changeover time is minuscule, the scheduler can potentially produce each part type daily. At this stage, there is less reliance on forecasting.

These are only some of the reasons why companies adopt the concept of structured flow manufacturing. The structured flow approach yields a variety of benefits, the rate of material utilisation increases, labour productivity improves, material transport time is reduced and scheduling is easier.

Costs of Structured Flow Manufacturing

The costs associated with implementing structured flow manufacturing tend to be facility costs, engineering costs and changes to administration systems. It involves redesigning the layout of factory equipment and altering operating procedures in the financial, human resource and information services departments to reflect these changes within production.

Facility costs depend on the extent to which equipment needs to be moved and the services which are already available. For example, heavy equipment may need reinforced floors, dirty processes may need extra specialised services, or electronics production equipment may need a clean room environment.

Engineering costs relate to dedicating resources to classifying parts into their various production families and allocating machines to these families. Engineering will also have to design the new layout and participate in implementation.

The information technology department may be providing services which are designed for a process-oriented environment. Structured flow manufacturing generally requires changes to software to enable it to monitor inventory and support production scheduling and finance.

The finance department may also have major concerns regarding allocation of overheads and valuation of inventory for their records. In process-oriented production systems, overheads are frequently allocated to the various machining areas as a percentage. Within the structured flow system, these costs will have to be allocated to the mini-lines.

The human resource department may have to re-evaluate the pay system to encourage operators to develop multiple skills and potentially to take on new responsibilities. Management responsibilities change from responsibilities for machining area to total responsibility for mini-lines. New performance metrics will thus have to be defined to reflect changes in priority for managers.

Costs are also incurred in lost production capacity during the implementation phase, which may last one or two weeks, depending on the scale involved. Any problems during restarting of equipment could potentially extend this period.

The many benefits and costs associated with structured flow need to be compared to decide the effectiveness of the solution. However, some of the benefits, such as improved customer satisfaction or employee morale, may be hard to estimate. Group technology could also yield information which enables the company to rationalise their product offering. In the machining industry, analysis may show that parts of similar dimensions and material, such as shafts, bushings and spacers, may be interchangeable, in which case the product offering can be rationalised without compromising the service provided. The ease in scheduling and its improved effectiveness is also difficult to gauge. A pilot introduction may be the only means of effectively measuring these benefits.

Analysis Work

Group Technology

The first step in designing structured flow manufacturing systems is to group parts according to their production requirements. There are three widely used sorting techniques:

1. Tacit judgement

2. Process flow analysis (PFA)

3. Standard codes/classification systems.

Tacit judgement is used when the variety of parts produced is low and their classification into groups is obvious. This is often the case in many industries, such as electronic assembly factories where printed circuit boards are assembled. They are initially grouped according to the level of technology required to produce them and then subdivided by the size of the board.

Process flow analysis (PFA) involves grouping parts according to the sequence of operations required for production. Figure 5.7 shows how parts are initially grouped according to the first operation required and sub-grouped according to subsequent operations, as applied to a machine tool factory. PFA is easy to understand and apply. The technique works best when the variety of parts is below 2,000. Companies using a greater variety of parts require a classification system which is faster to use and more descriptive of the part in question.

Companies with over 2,000 part types usually use numerical codes to classify parts into families. A typical code uses four to 25 digits to describe a part. Each number describes the answer to a question on a scale of 0–9. For example, the first number may describe whether the part is rotational or non-rotational, the second digit may describe the length of cut, the third digit may describe an external dimension or the number of surfaces for machining, and so on.

FIGURE 5.7(A): PROCESS FLOW ANALYSIS CHART: PRODUCT FLOW ANALYSIS

Product/Process	Lathe	Mill	Drill	Form	Assembly	Paint	Pack	Ship
a	1	2	3					4
b		1		2	3	4	5	6
c			1	2	3	4	5	
d	1	2	3					4
e			1	2	3	4	5	6
f	1	2						3
g		1		2		3	4	5
h		1			2	3	4	5

FIGURE 5.7(B): PROCESS FLOW ANALYSIS CHART: PRODUCTS WITH SIMILAR FLOWS GROUPED TOGETHER

Product/Process	Lathe	Mill	Drill	Form	Assembly	Paint	Pack	Ship
a	1	2	3	----	----	----	---- →	4
d	1	2	3	----	----	----	---- →	4
f	1	2	----	----	----	----	---- →	3
b		1	----	→2	3	4	5	6
g		1	----	→2	----	→3	4	5
h		1	----	----	→2	3	4	5
e			1	2	3	4	5	6
c				1	2	3	4	5

Computer-aided process planning software exists which acts as a database, holding this information and much more. The processing plan can be saved against the part and retrieved the next time it is required. When a new part is being introduced, it can be coded and processing plans for similar coded parts analysed to facilitate introduction. The person coding may discover that a part suitable for the customer's needs is already on the system. The system also aids estimation of costs since the estimator can look up the cost of producing similar parts.

Codes enable up to 150 parts to be coded per day. This is important in companies producing over 10,000 part types. The coding system relates to the production processes required to manufacture the part. Machines are grouped together to cater for a range of part classification numbers, or a computer algorithm can be written to perform the allocation of machines, as described in process flow analysis. Equipment is laid out in the sequence required to produce each family of parts.

However, the move to structured flow manufacturing can be restricted by a dependence on limited equipment capacity or availability.

Many manufacturing facilities have invested heavily in high speed equipment with the aim of processing a wide variety of parts through a single machine. For example, one multinational I worked with invested over £200,000 in a highly flexible solderwave machine and planned to process product from a variety of lines through this single machine. From a cost perspective, the machine was justified under the assumption that economies of scale in its usage made it a good investment. However, it made production control and scheduling much more complex, as inventories from a variety of areas had to be scheduled through the machine. Since there was only one of these machines, any disruption to its performance, such as downtime or quality problems, had repercussions for the rest of the production area.

An alternative to purchasing the multipurpose, high-volume, capital-intensive machine was to invest instead in a number of smaller machines so that one can be dedicated to the production of each family of parts. Less expensive machines were available but had less production options. Their lower cost meant that each cell could have had a dedicated machine, minimising time lost to equipment changeover and exposure to line disruptions and eliminating scheduling problems associated with multipurpose equipment.

Housekeeping and Five "S"

Any project involving a redesign of equipment layout should place strong emphasis on good housekeeping before making the move. Poor housekeeping can delay problems from surfacing and make the design of an effective layout more difficult. The area should be cleaned up before making the move, as otherwise the new layout will look dirty and confused. Many new layouts have been designed without good housekeeping in mind. As a result, redundant artefacts can be left scattered around, equipment cables can be left dangerously exposed in the way of operators, and operators themselves may be left with no place to store their tools. An effective housekeeping programme enables the layout engineer to understand and plan for the critical storage, space and facility requirements.

Good housekeeping is also critical for the safe operation of the whole production area. A disorganised work environment provides plenty of opportunity for oil spills, stray cables and unnoticed frayed wires. Good housekeeping also improves productivity, as tools are in their allocated places, so time spent searching for them is eliminated.

The Japanese have developed a five-step approach to workplace cleanliness. It is known as the "five S's":

1. *Seiri*: sort through and sort out.
2. *Seiton*: allocate storage, set limits
3. *Seiso*: shine the equipment and tools
4. *Seiketsu*: share information and standards
5. *Shitsuke*: stick to the rules.

The first S, *Seiri*, identifies what is needed and what is not needed for each operation. Items that are not needed are either discarded or stored in a separate area, perhaps a common storage area. Each person working on the factory floor should be able to justify all tools and equipment at their workstation. A guiding rule is that tools must be used daily in order to justify storing them at the workstation. This principle helps ensure that only essential materials are stored on the factory floor. Storage arrangements will be needed for tools that can be stored away from the workstation.

The second S, *Seiton*, defines specific storage locations for all tools and paperwork. The location is defined and sized to hold the recommended quantity of tools. Ideally, the storage location should be designed so that, by looking at it, the tools which are in use are readily identifiable. This can be done, for example, with wrenches by storing them in order of size, or by using a colour for each tool. The use of colour and the sequencing of tools makes the absence of a tool visible from a distance.

The third S, *Seiso*, simply involves shining the workplace. All equipment is cleaned thoroughly. A schedule for daily regular cleaning is set up. Cleaning may include doing simple preventative maintenance work around the machines.

The fourth S, *Seiketsu*, standardises the previous three stages and creates an open environment whereby information is readily communicated. In a standardised environment, problems are more easily identifiable. When a supervisor or manager walks through the area, problems such as build-up of inventory or a machine breakdown are quickly noticed.

The fifth S, *Shitsuke*, aims to engrain the system into the work culture. Unless focus is maintained on the first four Ss, the workplace can decline into its earlier condition. A system for continuously monitoring and auditing the workplace against the good housekeeping criteria has to be established.

Good housekeeping is recommended for any workplace but is especially important before any revision of equipment layout. It removes much of the confusion within the workplace and making planning for the redesign easier.

The Layout

The idea of structured flow manufacturing is to arrange equipment which produces similar parts in close proximity to each other so that transportation distance is reduced and operators take responsibility for the end product. However, there are advantages to the process-oriented layout which may leave management reluctant to experiment with structured flow manufacturing. The-process oriented layout results in lower facility costs. It also means that new operators receive a lot of support from colleagues operating similar machines. Specialisation in one machine makes the new operator very proficient.

FIGURE 5.8: STRUCTURED FLOW WITH EQUIPMENT GROUPED

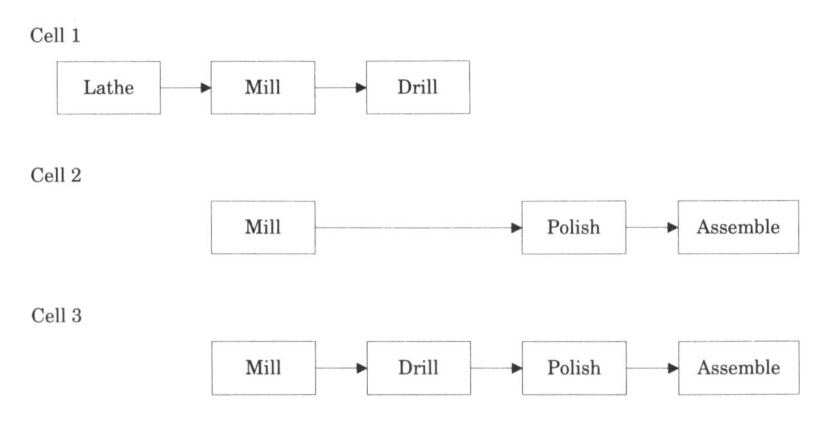

Cell 1

Lathe → Mill → Drill

Cell 2

Mill → Polish → Assemble

Cell 3

Mill → Drill → Polish → Assemble

Many of the advantages of process-oriented layouts can, however, be built into the design of the mini-layout. Figure 5.8 shows how a process-oriented layout has been converted to support the aims of structured flow. Facilities are localised and new operators receive support. Transportation distance is reduced, but the distance between some operations means it is not minimised. The layout can be seen as a compromise between idealistic structured flow and process-oriented layouts.

An ideal structured flow layout brings equipment into closer proximity and allows operators to move from one machine to the next in order to keep inventory moving through the mini-line. The U-shaped layout enables these two goals to be achieved. Figure 5.9 shows how seven machines are organised into a U-shaped layout.

FIGURE 5.9: U-SHAPED LAYOUT

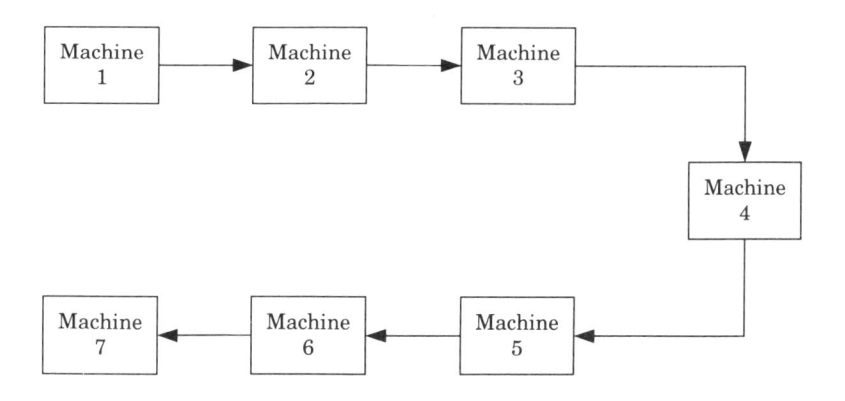

```
┌─────────┐    ┌─────────┐    ┌─────────┐
│ Machine │───▶│ Machine │───▶│ Machine │
│    1    │    │    2    │    │    3    │
└─────────┘    └─────────┘    └─────────┘
                                    │
                                    ▼
                              ┌─────────┐
                              │ Machine │
                              │    4    │
                              └─────────┘
                                    │
┌─────────┐    ┌─────────┐    ┌─────────┐
│ Machine │◀───│ Machine │◀───│ Machine │
│    7    │    │    6    │    │    5    │
└─────────┘    └─────────┘    └─────────┘
```

The advantage of the U-shaped layout is that it brings all the resources required to make the part into very close proximity. Operators, equipment, inventory, supervisors and engineers work together to produce the end part. Operators are multi-skilled so that they operate a variety of equipment within the area. At a minimum, each operator should be able to operate the machines immediately upstream and downstream. In this way, the operator is able to support their immediate colleagues. It also provides the operator with an appreciation of the difficulties their colleagues may have in operating their equipment.

Before finalising the layout, the implementation team should ensure that:

- Sufficient storage space is allocated for equipment and tools to be stored at each location;
- Overhead clearance and floor loading is suitable for the equipment being positioned at each site;
- Column spacing and positioning has been accounted for in the layout design;
- Water, drains, electrical power, air and vacuum are available as needed at new locations for equipment;
- Ventilation and extraction systems are capable of supporting the equipment;
- Aisleways and exits are in accordance with fire regulations;
- Adequate space has been provided for operating and maintaining equipment;
- The layout is in accordance with health and safety regulation.

A good layout enables product to flow smoothly from receipt, through the process, and on to shipping. It provides a comfortable and safe environment in which to work and reduces travel distance for inventory.

Balancing the Workload

A balanced production line means that work is evenly distributed between operators within the line. The same techniques can be applied to the mini-lines in structured flow manufacturing. The tasks required to produce a part are equally distributed amongst operators and ideally there is little or no need for inventory buffers between operations. As one operator completes their work on the part, the subsequent operator is ready to begin their work.

Just-in-Time manufacturing does not design lines to produce according to their maximum potential; rather it concentrates on having a line produce to maximum expected customer demand. For example, a line may be able to produce 120 units per day, but if the maximum expected demand for a two-year period is only 80 units per day, Just-in-Time distributes the workload involved in producing 80 units in order to balance the line.

A standard industrial engineering time study establishes the quantity of work involved in producing product from the beginning of the mini-production line to the end. In our example, 80 units are required per day on a single shift operation. The work involved in producing the 80 units has to be distributed between the correct number of operators and equipment.

The shift consists of eight hours of production. Therefore the maximum expected production rate is 10 units per hour. The mini-line will be designed to operate at a maximum of 10 units per hour over the next two years.

A work study reveals that each unit requires 0.5 labour hours to build. Therefore five operators (10 x 0.5 = 5) are required each hour. Theoretically, the mini-line needs five operators to perform at its maximum expected level of ten units per hour. However, since the mini-line uses equipment as opposed to being a purely manual line, the work cannot be distributed as simply as the above equation suggests.

The following table shows how the work is distributed amongst the various workstations/machines.

TABLE 5.2: DISTRIBUTION OF WORKLOAD FOR ONE UNIT

Machine	Workload	Req. Rate	Theoretical Operators Required	Spare time available with one operator
1	0.14	10	1.4	−0.4
2	0.08	10	0.8	+0.2
3	0.12	10	1.2	−0.2
4	0.16	10	1.6	−0.6
Total	0.5	−	5	−1

In this example, the workload is distributed among the four machines, as presented in the table (notice that the total is 0.5 hours, the number of labour hours required to make one unit).

Given that the required rate is 10 units per hour, this gives a theoretical number of operators required for each machine as 1.4, 0.8, 1.2 and 1.6 respectively for machines 1 to 4. — implying a total of five operators.

Clearly, with four machines, five operators are required to achieve the production rate; how the work is divided up depends very much on limitations within the cell and the layout of workstations relative to each other. The fifth and final column in the table details the amount of spare time or lack of time at each machine based on one operator "running" each machine. One example may be to have the fifth operator support the first and last workstations — as theoretically these require the highest workloads — thereby reducing their workload to 0.1 hours each per unit. Then the operator at workstation 2 could support the operator at workstation 3 (since the requirement at machine 2 is only for 0.8 of an operator, and the requirement at machine 3 is for 1.2 operators, the workload is more uniformly balanced). Supporting operators can help by performing tasks such as taking measurements for statistical process improvement, by cleaning their colleague's machine or by covering for breaks.

FIGURE 5.10: THE BALANCED U-SHAPED LINE

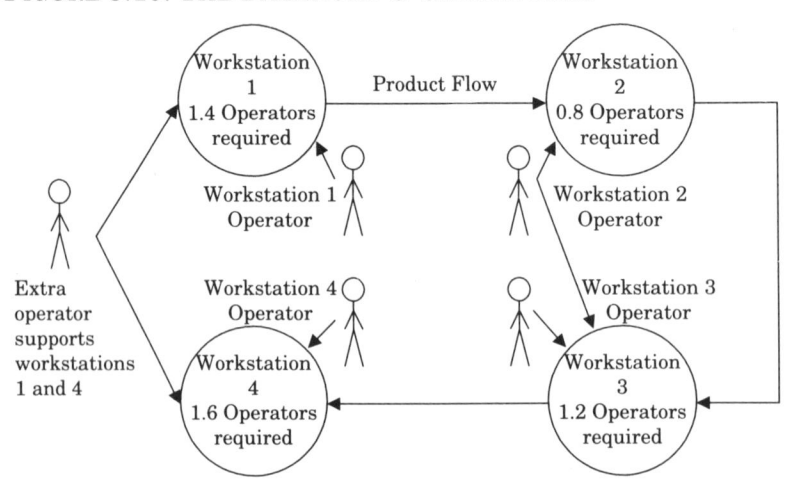

The U-shaped layout design is ideal for bringing operators within sufficiently close proximity to enable them to support their colleagues. This mutual support enables the mini-line to work as a close-knit team producing product at a constant balanced rate.

The mini-line has been designed to cope with 80 units per shift, and sufficient space, services and resources are available to support five operators. But what happens when only 64 units are required per shift? The calculation is performed at regular intervals, perhaps every month or every week, and people are added to or subtracted from the mini-line as dictated by demand. In the situation where only 64 units are needed per day (8 per hour), the calculation is as follows:

Machine	Workload	Req. Rate	Theoretical Operators Required	Spare time available with one operator
1	0.14	8	1.12	−0.12
2	0.08	8	0.64	+0.36
3	0.12	8	0.96	+0.04
4	0.16	8	1.28	−0.28
Total	0.5		4	0

This calculation is easily performed on a spreadsheet. In this situation, operators 2 and 3 support operators 1 and 4. Operator 3 may cover operator 4 for lunch-break, while operator 2 supports colleagues 1 and 4 during production. Operator 5 can be re-assigned to another mini-line or production area. The U-shaped layout and multi-skilled operators allow for this type of work distribution.

This concept of multi-skilled operators is a departure from the traditional specialist approach, where operators were encouraged to concentrate on performing only one task. The old school of thought was that productivity is gained when operators become specialised at one task. But proficiency in typical operator jobs can be achieved within 3–6 months. Therefore, operators should be capable of becoming multi-skilled and ideally able to efficiently work at each workstation in the cell.

Operators gain a better understanding of the level of quality they should be receiving and the effect of their workmanship by becoming familiar with subsequent operations. The work redistribution and the skill diversification programme provides greater continuity throughout the line and enables a linear outflow of product.

When a mini-line produces more than one product, the quantity of work must take into account equipment set-up time involved in changing from product A to product B. The work content now depends on the number of changeovers the schedulers plan for the month. Long equipment set-up time, especially on bottleneck equipment, can result in low productivity levels for the mini-lines. In general, production controllers attempt to reduce the number of changeovers in order to improve throughput. This approach reduces their flexibility in scheduling and increases their dependence on forecasts of expected demand. The Just-in-Time approach to resolving this problem is to effectively reduce equipment set-up time.

Reducing Equipment Set-Up

JIT aims to minimise the time and effort required to change over equipment from producing different product types. There are two reasons for this: first, it facilitates a balanced distribution of work as outlined in the section on balancing the workload; and second, the added flexibility it provides production scheduling reduces their dependence on forecasts.

In a situation where equipment changeover time is high, schedulers attempt to minimise the number of changeovers performed during the month or week. In a situation where product A represents 60 per cent of the required volume and product B represents 40 per cent, the scheduler may attempt to produce all the month's requirement of product A during the first 60 per cent of the month and concentrate on product B for the remaining part of the month. The scheduler depends on accurate forecasts to allocate production time. Finished goods of product A are stored as inventory, ready for the customer's order towards the end of the month. However, any changes in customer demands during the month can be difficult to accommodate when the factory builds to a forecast.

Just-in-Time manufacturing aims to produce according to firm customer orders rather than forecasts. This means that the changeover time between products should ideally be minuscule. Short equipment set-up times mean that the mini-lines can produce some of each product on a daily basis against firm customer orders.

If the equipment set-up time in the mini-line producing product A and product B is halved, then the schedulers can plan for an extra product changeover per month without losing any extra production time. When the changeover time is reduced by 75 per cent, the scheduler can organise a changeover each week. Ideally all equipment changeovers should be less than ten minutes, thereby enabling various products to be scheduled for daily production.

The expected number of changeovers to be performed can be factored into the workload for each workstation and the mini-line balanced accordingly. The reduced time involved in changeovers also means that variation in the number of changeovers performed form day-to-day will have less effect on the team's performance.

Figure 5.11(a) shows how the month's requirement for product A is built in one continuous period of time, and product B is built in a second continuous period. This means that the production schedulers must predict the factory's requirement for product A at the beginning of the month. Any inaccuracies in the forecast means that either insufficient or excess quantities of product A are produced.

This minimises time lost due to equipment changeovers. If an equipment set-up time reduction programme successfully reduces set-up time by 75 per cent, then product changeovers can be more frequent, without excessive loss of time. Figure 5.11(b) shows how the production scheduler can now plan production to a one-week forecast and can change between products more easily.

FIGURE 5.11(A): PRODUCTION BUILD BEFORE CHANGEOVER REDUCTION

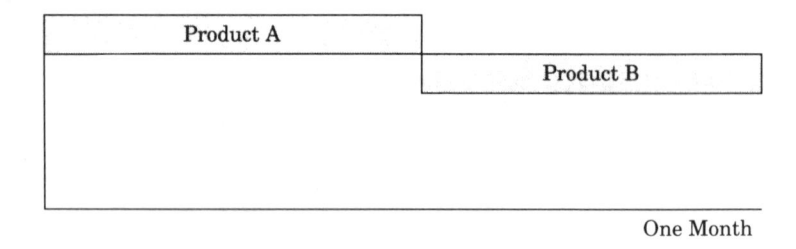

Product A	
	Product B

One Month

FIGURE 5.11(B): PRODUCTION BUILD AFTER CHANGEOVER REDUCTION

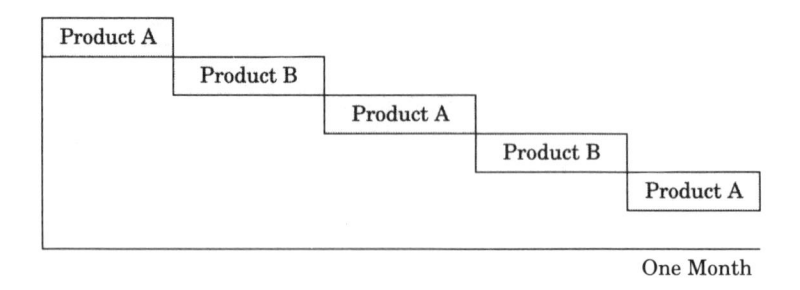

Product A				
	Product B			
		Product A		
			Product B	
				Product A

One Month

The first step recommended as part of a set-up reduction project is to video record a changeover in progress. A simple cam recorder with a timer will reveal much information about the changeover.

The second step is to analyse the video and list the tasks involved in the changeover.

The third step is to separate the tasks into *external* — i.e. those which can be performed while the machine is still running — and *internal* — i.e. those which necessitate stopping the machine while they are performed.

The fourth step involves generating ideas to convert as many internal tasks as possible to external tasks. This will result in the machine being down for less time. For example, on a printing machine, the operator might decide to purchase a second print wheel which would be set up, ready for production before beginning the changeover. With only one print wheel, the operator had to set up the print wheel in the middle of the changeover.

The fifth step is to minimise the time required to perform both internal and external tasks. This involves generating labour-saving ideas. For example, by networking electronic control systems on a machine, computer programs can be downloaded from a central terminal, thereby avoiding the need to change the systems individually.

The sixth step is to continuously reduce equipment changeover time, until zero set-up time is achieved.

Implementation Recommendations and Pitfalls

Each of the tools described above aids in the conversion of a process-oriented system to a structured flow manufacturing system. The move to structured flow should lead to a more even distribution of production — that is, the area producing product at a more constant and regular pace throughout the month. The techniques of group technology, housekeeping, layout, balanced workload and equipment set-up reduction should enable the scheduler to plan a more level throughput for each day of the month. An analysis of the actual throughput per day will reveal the extent of success.

The following examples give an overview of how the techniques can be implemented. The descriptions may help the reader to plan for their own implementation.

Group technology is a powerful tool, especially with companies that offer subcontracting manufacturing services since they generally cater for a broad range of products. GT enables the manufacturer to keep track of product details. The support software is available off-the-shelf, or can be designed by customising a standard database. With customised databases, particular information of interest to the manufacturer can be maintained and manipulated. For example, an electronics manufacturer can note the number of components per board or the average number of leads per component and thereby derive the expected usage rate of solder and flux per board. The time needed on the automatic pick-and-place machine can also be calculated and with similar information, a quote for producing the part is quickly available. The database can also indicate the line set-up used for similar parts and thereby reduce the time spent introducing the new part.

Housekeeping is an area that many factories perform poorly in. Employees focus on achieving production targets but put little effort into housekeeping. One of the reasons for this may be that there is no time allocated specifically to housekeeping. Some factories which emphasise housekeeping allocate the last five minutes of every shift for housekeeping only. The plant manager in a factory in which I once worked recorded parts of the factory on video, including the production lines and the offices. All staff employees were invited to attend the first showing. It was not pretty and much improvement was undertaken before the next video was taken. Another plant manager took all his employees through the "five S" system. He nominated days for cleaning and inspected each area in accordance with the five Ss. In both examples, the emphasis on housekeeping came from top management. Housekeeping succeeds only when it is driven from the top.

Layout design for mini-lines can be problematic for a multitude of reasons. For example, extra services may be required with the result that machines cannot be placed in the middle of the floor. But hard work and creativity will overcome many of these problems. For example, a factory had a machine which:

1. Needed to be enclosed because of the dirt and dust the process generated;

2. Needed to be up against an outside wall to minimise the draw for extract and to minimise exposure to leaks in the extract pipe;

3. Required operators to wear boiler suits and safety glasses because of the dirty and dusty environment.

The factory underwent an expansion during which this machine was thrown out and replaced by two new machines. The equipment engineer sourced a manufacturer to design a new machine where the dirt and dust was retained within the machine's cabinet. In fact, the dirt and dust generation was sufficiently reduced to eliminate the need for the enclosure but was still sufficient to require the operator to wear safety glasses and a protective apron.

It was decided to organise the machines into a mini-line format which meant that they had to be placed in the middle of the production floor. The machine manufacturers were commissioned to design an extract piping system. The design allowed for automatic shutdown of the machines if the pressure sensors detected a leak in the extract piping, with the result that the machines could successfully and safely be positioned in the middle of the floor.

The final improvement came to these machines when the material used for processing was changed. Dust and dirt generation was reduced and the extract system became more effective in removing waste; as a result, the operator no longer needed to wear glasses or protective clothing. It is seldom that all of the improvements needed on a machine take place at the same time. In fact, all of the improvements as described above were implemented over a period of approximately three years.

Mini-lines can also be difficult to design when equipment is part of the production process of several lines. For example, a factory had three mini-lines using a common set of equipment. In this situation, an industrial engineering study was performed and the flooring layout designed. The oldest equipment was grouped into the mini-line with the lowest expected throughput requirements. This line was placed at the greatest distance from the common equipment. New equipment was grouped into two separated lines with close access to the common equipment. Figure 5.12 shows the new layout.

Design of effective layouts can be hampered by the limitations and/or availability of equipment. Problems such as those described may make implementation of structured flow difficult, but sustained improvement work over time will overcome these problems.

FIGURE 5.12: U-SHAPED FLOW LINES USING COMMON EQUIPMENT

Two cells configured to use common equipment and to allow operators to support more than one machine

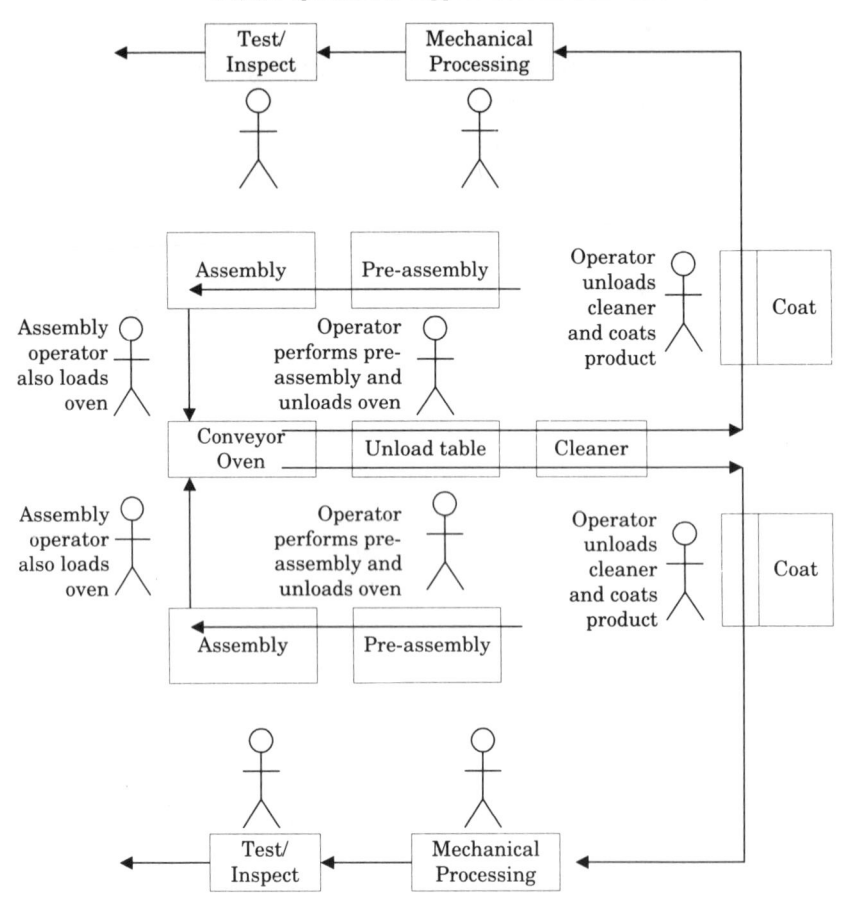

Balancing the workload can also be difficult. The workload was balanced for the mini-lines in Figure 5.12 so that when production throughput was low, the common equipment could be operated by operators working on the assembly equipment. When throughput increased, an extra operator was added to cater for common equipment and the work-load within each cell redistributed evenly.

Equipment set-up reduction is a powerful tool for improving the flexibility of machines. A factory with heavy-duty hydraulic presses performed a study to reduce machine set-up time from its current time of one shift. The process was videotaped and reviewed by a team comprised of industrial engineers, operators, technicians and the supervisor. The analytical steps for set-up reduction were performed and an improvement of 75 per cent was achieved. The main reasons for success were that the new die was heated up to production temperatures before commencing the changeover and the existing die was removed before being allowed to fully cool down. Other improvements stemmed from ensuring that a forklift truck and sufficient manpower were present before commencing the set-up. The clamping mechanism was also changed so wrenches were no longer required.

Each of the steps described moves the factory closer to structured flow manufacturing. However, the physical changes on the production floor need to be mirrored by similar changes in management structure. Supervisors change their responsibilities and objectives from machine utilisation in the process-oriented environment to product-oriented objectives such as product quality and on-time delivery.

THE KANBAN SYSTEM

The kanban system can be used to control inventory levels within mini-lines, or between mini-lines and the final assembly line. It is used to create a pull system whereby inventory is only produced as requested by operations downstream. The kanban system achieves these goals by using a tool — called a kanban — to achieve its objectives.

The word *kanban* is Japanese and translates as "card". Perhaps a more appropriate translation would be "request" because the word "request" indicates that a signal is being sent. With the kanban system, operations send signals or requests to their supplying operations requesting parts for processing. The supplying operations fulfil these requests by manufacturing parts and sending them to the originator of the request. By defining the request system, operations can be closely linked together and can mutually control each other's production.

This idea originated with Toyota executives observing the American retail system. They wanted to create a system where the action of the customer buying a product generated a chain of events which sent ripples through the supply chain, back to the factory manufacturing the part and even to the vendors who supply the factory. The purchasing action of the customer is considered as a request or kanban. The shop provides the part and requests a replacement part from the warehouse; similarly the warehouse requests a replacement from the supply factory, and so on.

Imagine a customer who has just purchased a new dishwasher from a hardware store. From the customer's perspective, the sales shop had the right machine in top quality condition at the right price. After the sale, the store manager probably telephones the warehouse to send a replacement machine to the shop. The replacement machine is delivered on the same day and again is in top quality condition and correctly priced.

In order to track the warehouse inventory, the warehouse manager has developed a system with the dishwasher manufacturer such that all dishwashers arrive with a special identity card. The warehouse manager ensures that his shipping staff remove and store the card before delivery to any shop. These cards will be used to request dishwashers from the factory and can therefore be considered kanban cards.

FIGURE 5.13: SHOP AND WAREHOUSE SUPPLY CHAIN

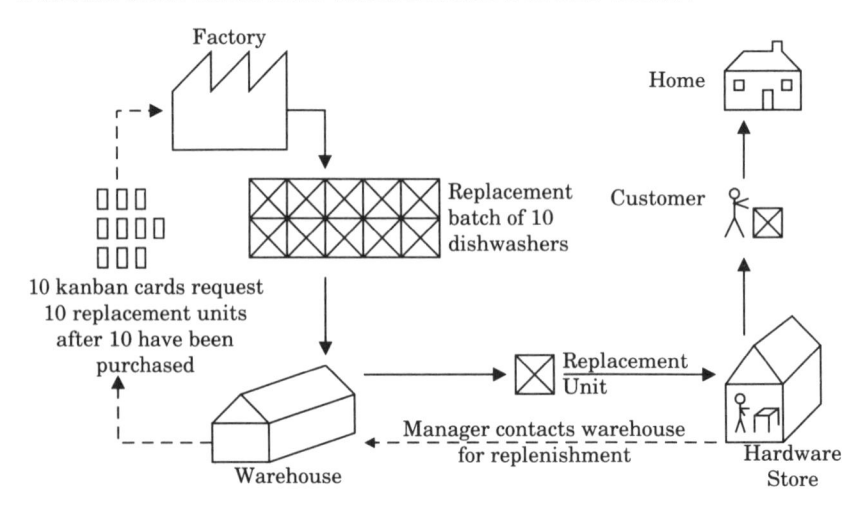

Perhaps after ten of these dishwashers have been sold, the warehouse requests ten replacements from the supplying factory. In a kanban operating environment, the warehouse would send ten kanban cards to the factory, each card requesting a dishwasher for replacement. The factory managers have trained their staff to consider the cards as purchase orders. When the dishwashers are ready, one card is attached to each machine. The dishwashers arrive at the warehouse with their kanban cards. As they are distributed to shops, the cards are removed from the packaging and stored by administration. When ten cards have accumulated, the administration clerk repeats the cycle with the supplying factory. The number of dishwashers in the warehouse or on request is defined by the number of kanban cards.

When the factory receives the ten kanban cards, they are sent to the final assembly line and ten machines are built to satisfy the order. The machines are packaged and the kanban card put on each machine. After these machines are built, if there are no more cards from any other source, the final assembly line stops producing. The final assembly line uses ten motors, ten cabinets, ten control systems, ten sets of hosing, etc., to make these ten dishwashers. All these items need to be replaced. Attached to each of these items is a kanban card. As the item is utilised, the card is removed and sent to the supplying line or operation as a signal for a replacement part to be supplied. The process is repeated down the supply chain, to upstream operations and to supplying factories. In our example, the motors are supplied by an external supplier. The kanban cards are sent to the supplying factory where, ideally, a similar chain of events occur in providing replacement orders.

JIT and kanban cards tend to be relatively easy to manage, once set up, but as the simple example above demonstrates, a lot of co-ordination and analysis is required to develop the system initially. The kanban is the communication signal that links operations and enables the supply chain to act as a single entity with the aim of minimising production and inventory stocks.

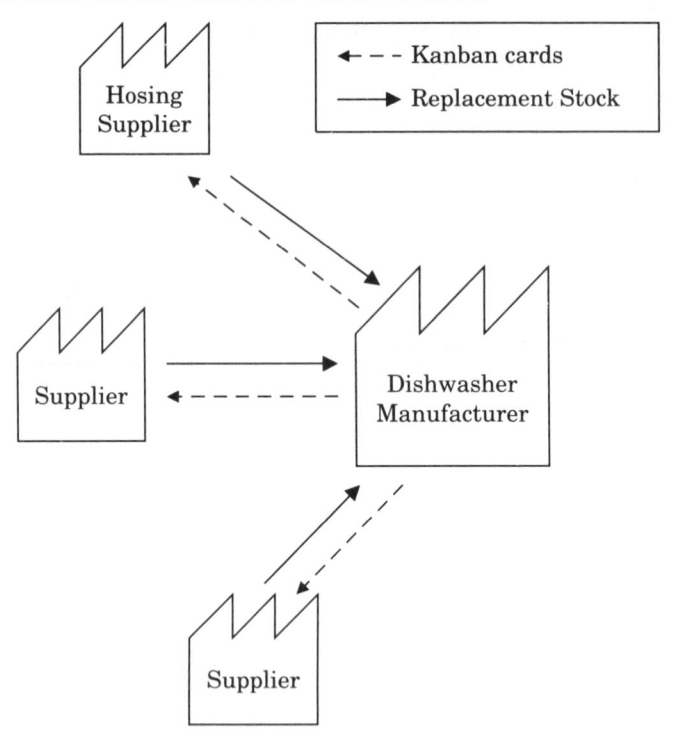

FIGURE 5.14: A FACTORY AND ITS SUPPLIERS

Implementation of the kanban system is generally easier when the supply chain performs predictably. For this reason, structured flow manufacturing and effective quality and equipment maintenance are recommended as forerunners to the introduction of a kanban system. The following model is proposed as a guide for programme managers to adopt or manipulate as needed. The model addresses the introduction of the kanban system onto the production floor. It consists of:

- *Preliminary work:* Evaluate costs and benefits.
- *Analysis work:*
    ◊ Define acceptable inventory levels
    ◊ Define the kanban mechanism
    ◊ Design one-piece flow.

- *Implementation recommendations:* General guidelines for implementation.

Similar to structured flow manufacturing, the kanban system affects a wide range of people working in the factory. It involves reducing batch or lot sizes for product and changing the way material is scheduled through the production area.

The easiest way to overcome these anxieties is to create a team with members from each sector of the company to investigate and determine how the kanban system can be introduced with the minimum disruption. That is not to say disruption is to be avoided. The overall goal is an efficient process that adds value and minimises waste. Some departments that depend on a process-oriented manufacturing environment will have to radically change their systems. Co-operation, teamwork, understanding and a commitment to success should overcome any problems.

Preliminary Work: Cost–Benefit Analysis

The cost of implementing a kanban system can vary. They can be low, where the main costs incurred lie in dedicating internal resources in designing and implementing the new system. Additional resources from external consultants may also be required but cost is easily controlled and budgeted for.

In contrast, costs can soar when the new system is poorly designed or implemented. It leads to lost earnings through lost production. It can generate confusion on the factory floor and more importantly can lead to loss of credibility and support for the implementation team. Careful design and pilot implementation reduces these risks.

Many benefits accrue from the kanban system. They include reduced and controlled inventory levels and consequently easier scheduling. One manager who observed a production area operating a kanban system professed his amazement that, over a period of months, the inventory in front of each machine never became excessive, yet never ran out. Since inventory levels are effectively controlled, manufacturing lead-time also improves, as indeed does the quality, since reduced inventory means quality problems are detected faster.

Analysis Work: Defining Acceptable Inventory Levels

In the idealistic JIT environment, the only inventory at a workstation is that which is being processed. In reality, buffer inventory is also present because it helps to smooth the production. The implementation team has to agree on acceptable inventory levels to be maintained between operations. The defined inventory levels should be small enough to ensure that the company achieves goals laid out in the preliminary analysis while they must be large enough to avoid the production area coming to a standstill every time a machine has downtime.

Traditional manufacturing management uses inventory between operations to protect machines from problems that occur in other areas of the production line. A common analogy used to describe the effect of inventory on problems is how water protects a boat from hitting rocks in a river bed (see Figure 5.15).

FIGURE 5.15: THE ANALOGY OF A BOAT PROTECTED FROM THE RIVERBED BY HIGH WATER

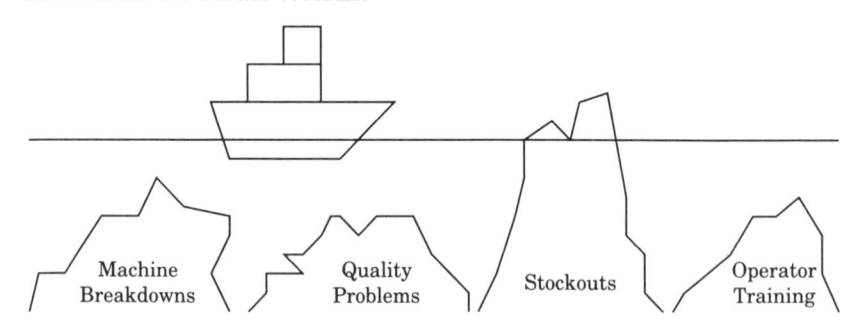

JIT reduces the water (inventory level) to expose the rocks (problem). Ideally as exposed problems become resolved, the inventory level is further reduced, exposing new challenges. These problems can then be continuously reduced by using techniques such as those described in Chapter 2.

Just-in-Time Production

The goal is to ensure the quantity of problems exposed is within the capabilities of production staff to efficiently resolve. If too many problems are exposed, the line grinds to a halt. If too few are exposed, it negates the expected benefits from reduced inventory. The team of production supervisors, maintenance, materials, quality engineering, process engineering and operators need to decide the level of exposure which can be coped with. The team sets the maximum permitted inventory levels between operations within the mini-line and between mini-lines. If this inventory level is about to be exceeded at any stage, the operation comes to a halt.

There are a number of variables other than random line disruptions that affect the optimum inventory size. These variables include the daily production rate, the lead-time for replacement of parts and the quantity of parts used per assembly.

Computer simulation, covered in Chapter 6 is recommended for the design of inventory buffers. It allows actual data and random effects to be modelled and measured. Various inventory sizes can be tested and their effects estimated. Experimenting with computer simulation is generally much cheaper than running pilot testing on the actual line.

FIGURE 5.16: KANBANS LINKING CELLS AND CONTROLLING INVENTORY

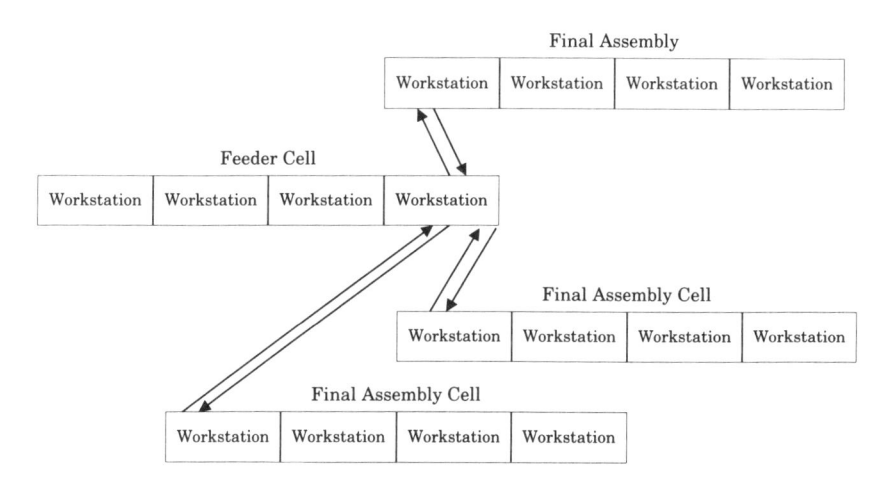

Once the maximum acceptable inventory buffers are established and operational, they can be continuously reduced in a controlled manner that gradually reveals new problems for engineering and production staff to resolve. As inventory levels are reduced and problems are resolved, the production area becomes much leaner and more efficient.

Defining the Kanban Communication System

The next step is to define the communication signal, the request system that controls the movement of inventory in and out of mini-lines and operations. The kanban communication system regulates the flow of product from the first to the last operation.

When an operation uses parts from its input buffer, it communicates to the previous operation what parts should be replenished. The kanban communication system dictates to each previous operation what, when and how many of a particular part is requested and therefore needs to be produced. In this way, units are "pulled" through the line depending on what is being used by the final operation.

In Japan, a card is generally used, describing to previous operations what needs to be produced in order to replenish the inventory buffer.

FIGURE 5.17: SAMPLE KANBAN CARD

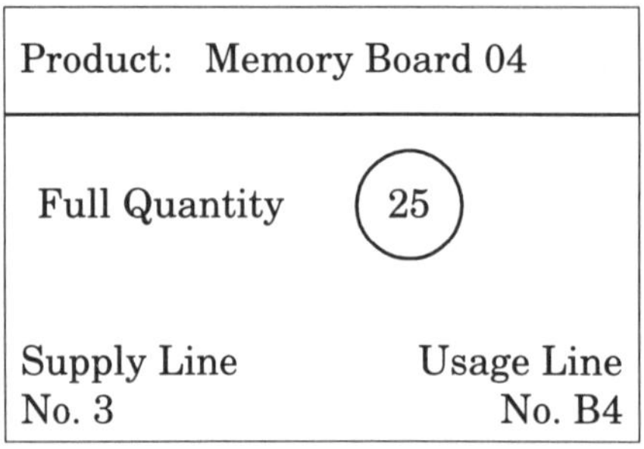

Product: Memory Board 04	
Full Quantity	25
Supply Line No. 3	Usage Line No. B4

There are a wide variety of tools that could be used to communicate this signal. The important criterion is that the system is easy to use and understand. In one company where I worked, we used standardised containers, a system with a number of

Just-in-Time Production

advantages. Physically they were only able to hold specific quantities of inventory. This meant that the maximum inventory levels were defined by the number of containers allowed within the system. They were labelled and supervisors were responsible for ensuring that only specified material was in the bin. When an empty bin was sent to an operation, it was a signal/request for product specified on the label. When no empty bins were available at an operation, the operation had reached its maximum allowed inventory and came to a standstill. The operator was then reallocated to other work. In a well-balanced line, which has few disruptions, this should be an infrequent occurrence.

Standardised containers are good at preventing "well-intentioned" but misguided operators producing product even when there is no request. It can be hard to reprimand, as this often demotivates an operator who thought they were doing what was best for the business. However, the standardised bins ensure that the operator contacts the supervisor regarding the lack of work. The lack of empty bins in which to place parts constrains the operator's output.

There are some simple rules which control the flow of kanbans and correspondingly control production.

1. Only parts specified by the kanban should be obtained. In our example, only the parts specified on the bin label should be stored in the bins;

2. People should only produce parts as specified by available kanbans. The kanban is the request; without a request for product, the operation stops;

3. Only quality parts should be made available for kanbans. The operator should take responsibility for the quality of the product being sent to the subsequent operation;

4. The kanban should always be attached/linked to its associated parts until the parts have been used up.

Kanban control systems operate automatically when these simple rules are adhered to. These rules apply equally to management as they do to operators. It is important that production schedulers and supervisors adhere to these guidelines and avoid bypassing the rules in order to expedite urgent orders.

The kanban communication system links all operations within the factory. The customer's demand on the final assembly operation is transferred throughout the factory through a series of kanban exchanges; as a result the whole production area begins to operate as a single entity. The system tends to be self-regulating which makes production control and scheduling much easier. As product is pulled from the final assembly line to fill customer orders, the supplying operations and mini-line automatically generate replacement units.

Large versus Small Lot Size and One-Piece Flow Production

One-piece flow means having a lot size of one unit. Traditionally, production systems move product for processing between operations in lots or batches of between tens of units and thousands of units. This makes tracking units through the production floor easier. In an area producing 500 units per day in lots of 50 units each, the production scheduler tracks the progress made with each 10 lots instead of 500 individual units.

Another advantage of large lot sizes is that it reduces the labour time involved in transporting units between operations. The next processing operation may be a significant distance away from the originating operation, especially with process-oriented layouts. Transferring 500 units individually requires more labour than moving 10 lots onto the next operation.

Large lots also minimise the set-up time on machines when, for example, there are a variety of product types included in the 500 units. Since the product moves in lots of 50, the operator gets a minimum of 50 units from the machine before having to change the machine settings. But the large lot size approach also has problems.

Large lot sizes result in increased lead-time for production, as demonstrated in Figure 5.18 which shows the time required to produce 10 lots of 50 units as opposed to 500 individual units. It is assumed in the Figure that there is no equipment set-up time required and transportation time is negligible. The diagram shows how producing in a lot size of one — i.e. one piece flow — significantly reduces production lead-time by minimising or eliminating delays within operations; that is the time delay which occurs in the lot of 50 units, 49 units having to wait at the operation while the other unit is being processed. The bigger the lot size, the larger this delay.

FIGURE 5.18: SMALL LOT SIZE REDUCES LEAD-TIME

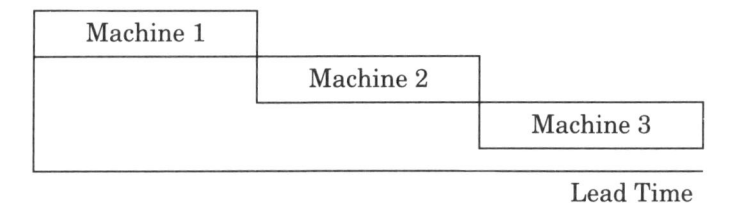

```
| Machine 1 |
       | Machine 2 |
              | Machine 3 |
                    Lead Time
```
(a) Time to move a large lot of 500 through three different processes

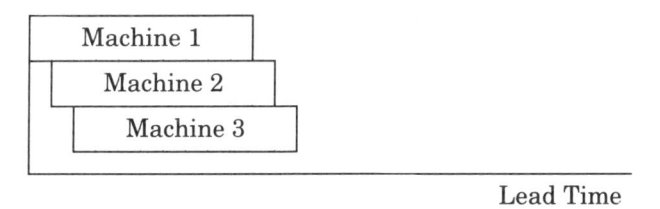

```
| Machine 1 |
 | Machine 2 |
  | Machine 3 |
         Lead Time
```
(b) Time to move 10 lots of size 50 through three different processes

One of the problems regarding one-piece flow relates to tracking product through the floor. There is no simple solution to resolving this problem. Much of the software for inventory tracking systems was developed for a Material Requirement Planning (MRP) environment where material is moved in relatively large lots. The large lot production method is easier to track, but one-piece flow does not mean that 500 units will have to be tracked individually. Units can be in one of three significant locations:

- The input buffer to the cell;
- Being processed within the cell; or
- The output buffer after processing.

One means of aiding tracking is to identify each unit with a serial number. When the number of the next unit to enter the production line and the last unit processed are known, then the general whereabouts of all units can be derived (see Figure 5.19).

The nature of one-piece flow means that the inventory level within the process is low and the speed of units through the cell is fast. The low inventory and fast processing speed often negates the need to further break down the position of inventory within the process. Since processing time is quantified in days rather than weeks, the location of all batches are easily tracked.

FIGURE 5.19: ONE-PIECE FLOW AND SERIALISATION OF UNITS FOR TRACEABILITY

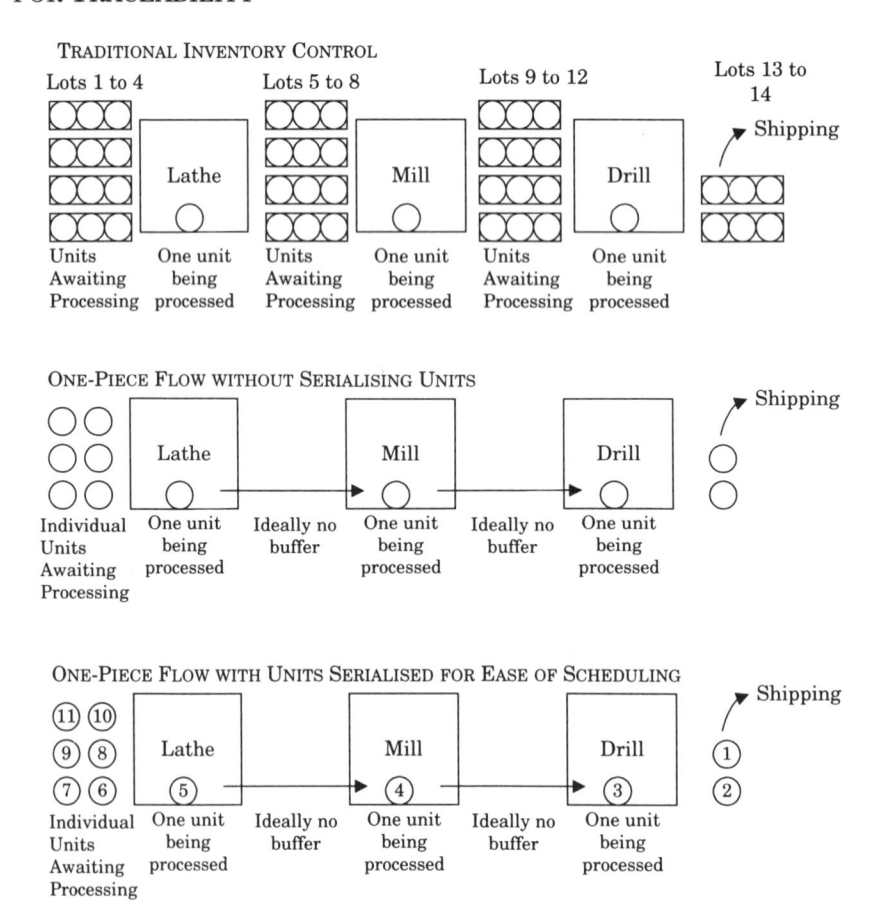

TRADITIONAL INVENTORY CONTROL

[Diagram showing Lathe, Mill, Drill stations with Lots 1 to 4, Lots 5 to 8, Lots 9 to 12, Lots 13 to 14 shipping — Units Awaiting Processing / One unit being processed at each station]

ONE-PIECE FLOW WITHOUT SERIALISING UNITS

[Diagram showing Lathe, Mill, Drill with Individual Units Awaiting Processing, One unit being processed, Ideally no buffer between stations, Shipping]

ONE-PIECE FLOW WITH UNITS SERIALISED FOR EASE OF SCHEDULING

[Diagram showing Lathe, Mill, Drill with numbered units 11, 10, 9, 8, 7, 6 awaiting; 5 being processed at Lathe; 4 at Mill; 3 at Drill; 1, 2 awaiting Shipping]

Many of the other obstacles to implementing one-piece flow are resolved by the structured flow manufacturing system. For example, the travel distance of small-lot production must be short between machines. Ideally, to set up a one-piece flow system, it is necessary to minimise transportation time between operations to practically zero. This is achieved within the mini-line layouts of structured flow manufacturing. The close proximity between operations means that when one operator leaves down the processed unit, it is within easy reach of the next operator.

Equipment changeover time between parts must also be very low to support small-lot production. The equipment set-up reduction programmes described earlier ensure that the minimum time is lost to product changeovers. Reduced set-up means that the product mix going through the area can be varied without causing excess production time loss.

Small-lot production significantly reduces manufacturing lead-time. It minimises time delay within operations, because when a unit is being processed, all the other units within the lot wait. As described earlier, in a batch of 100 units, 99 must wait while the 1 is being processed. When the batch size is halved, only 49 wait as each unit is being processed. Small lot production reduces manufacturing lead-time and improves the responsiveness of the factory.

Implementation and Recommendations

A pilot trial of the kanban system is strongly recommended to minimise exposure to production loss. A pilot introduction gives operators an opportunity to become aware of the rules associated with the new system and an opportunity to participate in the development of the system.

A kanban system need not be introduced between all operations simultaneously. One implementation team could install their kanban system for operations upstream from the bottleneck operation. In this way the pilot implementation effectively controls inventory for the first section of production while exposing throughput to the minimum risk.

On a multi-shift operation, it is advisable to design forms which supervisors can complete if they deviate from the rules of kanban. On off-shifts, there may be little support available, and stopping production can be an intimidating responsibility. If the rules are broken, a questionnaire describing the situation will enable the implementation team to consider the event and clarify what the appropriate action should have been.

It is very important that the whereabouts of kanbans be known. It is advisable to check this regularly because if kanbans are lost, acceptable inventory levels are misrepresented.

The following example highlights many of these points. A manufacturer wanted to reduce the processing lot size in order to reduce the production lead-time. The line consisted of 9 sections; however, only the beginning of the line is described here as it is sufficient to clarify the changes made and benefits derived. This part of the line consisted of assembly, firing, encapsulation and cleaning; Figure 5.20 shows how the area was laid out. An ideal cell layout was not possible, because the encapsulation machines had to be enclosed in a confined area for environmental reasons. The diagram shows that, even though the machines are not in close proximity, the implementation team purchased a conveyor to overcome distance and establish a much reduced lot size.

FIGURE 5.20: LAYOUT OF AREA

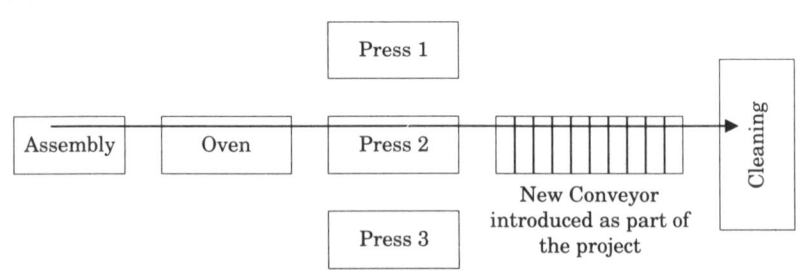

With the old production control system, lot sizes varied between 4,000 units, the minimum order quantity, to 150,000 units. When a customer ordered over 200,000 units, the order was broken up into lots of 20,000 to help the product move through the floor in manageable quantities. The line produced several million units per month. In this high-volume environment, any attempt to handle each individual unit separately would be impracticable and unnecessary considering the minimum order quantity of 4,000 units. It was decided to concentrate on achieving a transportation quantity of 1,000 units, i.e. that 1,000 units would be produced and moved on to the next operation before processing the next 1,000 units.

The daily production rate was calculated and the work content for each operation was calculated and distributed. It was decided to label each set of 1,000 units with a serial number and operators were instructed to process all sets sequentially. This meant that the transportation quantities did not get mixed up in a flood of work-in-progress. Inventory between operations was controlled using kanbans. The new system was set up on a trial basis and it was soon identified that the encapsulation operators were spending too much time transporting the transport quantities. A conveyor was installed to reduce this transportation time and encapsulation was allowed to work with transportation quantities of 2,000 instead of 1,000 in order to reduce material handling and transportation time.

This system reduced the production lead-time by 50 per cent; once the conveyor was installed, the transportation time was reduced and productivity improved. With the old system, the large customer orders were broken into maximum lots of 20,000. This resulted in extra administration work, as lots were set up on the computer and work sheets had to be filled out. The new system allowed a large lot such as 100,000 to progress through the production floor in transportation quantities of 1,000 and 2,000. Administration work was only recorded for the 100,000 lot and not for the individual 1,000–2,000 transportation quantities. This significantly improved operator and production control productivity.

Production control personnel made significant gains from this new production system. They were guaranteed that units came out of the line sequentially, which meant that instead of having to prioritise work for each operation, they prioritised work entering the line. Also, there were less customer orders to be tracked within the line and the production lead-time was significantly reduced.

The small-lot production system was introduced at the same time as kanbans for inventory control. However, kanbans can cause line stoppages which result in reduced capacity when the constraining (bottleneck) operation has to stop. The advantage in this example of introducing both systems simultaneously was that the productivity gains made from moving to smaller lots easily compensated for any initial capacity loss due to the kanban system. It is important to realise that, over time, the kanban system tends to more than recuperate any initial loss in capacity. This is because the system tends to pressurise the production process, resulting in much focus being placed on eliminating the causes of line stoppages.

JUST-IN-TIME EXTERNAL SUPPLIES

JIT deliveries are the third component of an overall JIT system and involve suppliers delivering raw materials just as they are required on the production floor. Regular and frequent deliveries from suppliers reduce the overall manufacturing lead-time by reducing the time raw materials spend in storage. JIT deliveries mean:

- Working capital invested in raw materials is reduced;
- Warehousing requirements are reduced;
- The potential for goods being damaged in storage is eliminated;
- Reduced economic danger of raw materials becoming obsolete due to product design changes or new product introduction;
- Any defects in raw materials are quickly noticed because they are processed soon after delivery. Fast feedback to the supplier on quality issues improves the overall quality of raw materials.

Three steps are outlined for setting up JIT deliveries. The preliminary work involves identifying which raw materials should be delivered on a JIT basis. The second step is analysis work which establishes the means to enable JIT deliveries to be implemented. The third step, implementation, is illustrated with an example of how JIT delivery helped one company.

Preliminary Work

When a bill of materials contains thousands of parts, it may not be feasible or worthwhile to arrange JIT deliveries for all parts. In general, parts can be divided into two categories according to cost. By applying the 80:20 rule, it can be assumed that 80 per cent of the cost is associated with 20 per cent of the parts. These 20 per cent are the initial candidates for JIT deliveries. Typically, it is cheaper to store low-cost parts rather than undertake any extra transportation costs. There may be reasons other than cost driving the JIT delivery project; some of these may include:

- Bulky parts are expensive and difficult to store;
- Acids or caustic chemicals are dangerous and best maintained in minimum quantities;
- The parts are subject to many engineering changes and inventory must to be closely controlled.

In order to effectively introduce JIT deliveries for these parts, some fundamental conditions must be satisfied. The daily requirements of parts should be predictable, the parts supplied should be of a very high quality and the supplier should be capable of making frequent deliveries.

Analysis Work

The JIT delivery system differs from the traditional approach of supplier management. Traditionally, multinational companies have multiple suppliers for the same components in order to ensure price competition between suppliers and to protect the company from exposure to variations in supply or demand. This policy provides protection against an industrial strike at any individual supplier, or if there is a peak demand for product, the demand can be spread over a variety of suppliers. The negative side is that, since the customer generally represents only a fraction of the business of any supplier, their ability to influence the supplier's production system is reduced. Other disadvantages are that the customer's production system has to be robust enough to cater for raw materials from a variety of suppliers. This significantly increases the level of variation on the production floor and leads to variation in the quality of goods produced. Also, maintaining a broad supplier base leads to duplication of work for the purchasing department, since administration work increases as the number of suppliers increase.

In a JIT environment, the adversarial relationship between supplier and customer is replaced by a partnership relationship where mutual support replaces distrust. The greater level of trust is reflected in the quality of goods supplied. The customer expects the supplier to protect their business by supplying only top quality goods. Therefore, there should be no need for incoming inspection of raw materials. When reducing the number of suppliers, companies give preference to suppliers who produce high quality goods and operate in the locality. Both characteristics make the implementation of JIT much easier.

Suppliers who operate in close proximity to their customers will have the least logistical problems in arranging JIT deliveries. The short distance between supplier and customer also improves communication between personnel in both companies. Meetings can be easily arranged to discuss issues as they arise. There is also closer co-operation in the development of new products and, in addition, mutual support can be provided to resolve various problems, such as engineering, production scheduling and quality, as they arise.

Some suppliers who have their production facilities a significant distance from their customers, perhaps even overseas, have overcome the problem of proximity by delivering supplies from a local bonded warehouse. These companies supply the bonded warehouse in bulk and deliver to their customer from this site. They take ownership for the level of raw materials stored in the warehouse and if there are several customers in the locality using similar parts, the warehouse enables them to provide a better service. For example, a supplier of computer peripherals like monitors may set up a similar system and supply several computer companies from the same stock. The cost of the service can then be spread across a range of customers. The disadvantage of the warehouse system can be seen when it supplies only one customer; then, the customer must support the warehouse system as part of the product cost.

The quantity to be delivered will be a function of daily usage and of the frequency with which deliveries can be made. The success of the project will depend on the customer being able to predict relatively stable usage rate for goods. The supplier will need access to their customer's production scheduling reports and sales forecasts in order to plan their production to mirror that of their customer.

Implementation Guidelines

If deliveries are made directly to the point of use on the production floor, the implementation team needs to arrange suitable short-term storage facilities at these points. Procedures will have to be set up to ensure that a "first-in, first-out" (FIFO) usage system operates. With these arrangements in place, the supplier can replenish goods as they are used.

The following example describes how a company set up JIT deliveries for its packaging materials. The company was expanding and needed to assess the level of warehousing required to hold packaging material for its shipping department. Since packaging materials are, by their nature, bulky, the warehouse would have to be significantly increased with a corresponding increase in inventory carrying costs. The alternative was to implement JIT techniques to minimise the need for storing packaging materials. Senior management decided to implement a JIT delivery system. They communicated the need to their external suppliers and their operators in the packaging department. A team comprised of company and supplier staff was set up to design how the system would work.

A total of 2,500 boxes were allowed in the delivery pipeline. This quantity was represented by 25 kanban cards, each of which moved with 100 boxes from the supplier's shipping department to the customer's "point of use", and back to the supplier when the boxes had been utilised. The quantity in the pipeline could be further reduced if the supplier was able to deliver on a more frequent basis.

On a slow production day, the rate of box usage was reflected in the low number of kanban cards returned to the supplying company. On a high production day, the cards were returned faster. This simple system was self-regulating; each card was treated as an effective purchase order. The receiving department recorded all deliveries and the size of deliveries mirrored the rate of utilisation. Warehouse needs were minimised as the supplier held a buffer to compensate any mismatch between customer demand and rate of supply. In addition, the supplier was able to use this buffer to service a number of clients. The boxes were customised and printed only upon receiving a kanban card. The single buffer of boxes was efficiently maintained to cater for surges in demand from a pool of customers.

The system minimised the need to store packaging materials, and correspondingly controlled the level of damage to packaging inventory. Inventory carrying costs and indirect administration costs were reduced. It tied the supplier into a closer relationship with the customer whereby they were able to plan their own production level to correspond with their customer's production schedule.

CONCLUSION

Any programme to install a JIT manufacturing system has to address how the fundamental building blocks — structured flow manufacturing, the kanban system and JIT deliveries — will be implemented.

JIT concentrates on value-added material utilisation. The system moves raw materials quickly through each of the production processes and on to the customer. Ideally, as each workstation finishes processing a unit, the subsequent workstation prepares itself to begin its processing.

Just-in-Time Production

FIGURE 5.21: THE JIT SYSTEM

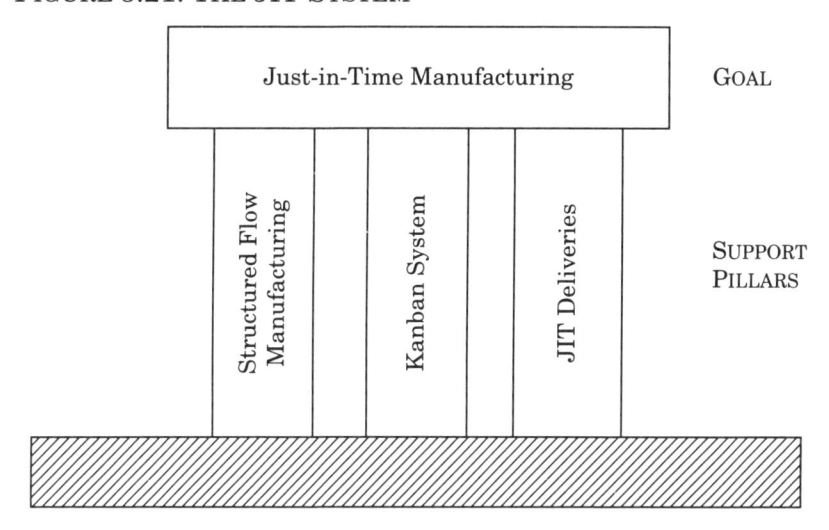

```
┌─────────────────────────────────────┐
│     Just-in-Time Manufacturing      │    GOAL
└──┬──────────┬──────────┬────────────┘
   │          │          │
   │ Structured│ Kanban  │ JIT        SUPPORT
   │ Flow      │ System  │ Deliveries PILLARS
   │ Manuf.    │         │
═══╧══════════╧══════════╧═══════════
```

JIT is a continuous improvement programme. Major benefits are typically achieved after the initial introduction phase. However, by continuously reducing inventory sizes to reveal further production problems, further progress can be made in productivity, waste elimination and lead-time reduction. The JIT approach simplifies manufacturing systems, since the shorter production lead-times mean that factories are less dependent on forecasts to dictate what to produce.

The JIT system is a revolution in production control techniques. It requires co-operation and innovative ideas from everybody working within the production area to overcome implementation problems. Inevitably, companies will have particular problems associated with their production techniques which make JIT implementation more difficult; however, a team approach to problem-solving should enable obstacles to be overcome.

Chapter 6

COMPUTER SIMULATION

INTRODUCTION

Managers are increasingly using computer simulation as a tool to aid decision-making, to help them cope with today's changing industrial environment. Computer simulation is used to model dynamic real-life systems such as factories and distribution systems, thus enabling managers to investigate proposed changes and to predict the effects of these changes on the system. It means that management can experiment and test their ideas before putting them into practice. Computer simulation allows management to confidently introduce change because they have already validated their ideas before implementation.

In the 1970s and the 1980s, computer simulation was difficult to learn. Simulation packages such as SLAM or GRASP required its users to understand a programming language such as FORTRAN. Simulation was associated with computer programmers who managed to digest lots of data and produce pages of results. In the late 1980s, simulation software became available which was easy to program, could be run on personal computers and was comparatively inexpensive. In the 1990s, animation capabilities were added to these packages so that the user could actually see the system operating on the computer screen and more easily explain to colleagues the system on view.

Computer simulation helps management to make more informed decisions. The manufacturing environment can be full of uncertainty. Random variations have knock-on effects over a range of areas. Variation in demand can affect the factory's performance against on-time delivery; or variation in machine performance can affect the level of inventory storage required. There are many examples of variables that affect a system and can cause a high degree of variation. When managers are faced with such complex problems, it may not be clear how they can best be resolved.

Static modelling tools such as spreadsheets do not accurately reflect these interdependencies. They describe system components through equations that are only calculated once, or use macro-subprograms to perform a series of routine calculations. Dynamic computer simulation adds a time dimension and calculates the state of the system as time passes. It effectively imitates the system by reacting to the passing of time and random events such as equipment downtime.

Computer models can be used on improvement projects for "on-time delivery" performance, for reducing manufacturing lead times, for improvement in capacity utilisation, or to calculate queuing times. They provide an insight into the problem and possible solutions.

The very nature of building a computer model involves gathering data and asking questions about the system. This work helps the users to gain an overall understanding of the system under examination. Much of the benefit associated with simulation comes from gaining new perspectives on how the system operates. It can save both time and money by predicting trouble areas and enabling management to confidently design and introduce change.

DEFINITION OF COMPUTER SIMULATION

Computer simulation is a mathematical representation of a system. The system is broken into its components and the behaviour and interaction of each of these components is mathematically modelled on the computer. The software packages are designed to facilitate ease of modelling. The end result is a computerised imitation of the real system. Just as a word processor simulates a typewriter and piece of paper, or an arcade game represents space invaders, computer simulation packages can also represent factory systems.

Modelling involves breaking the system down into its component parts and defining each component's individual characteristics. Modern software packages offer pre-built modules to represent many of the most common components of a system; for example, in the factory context, the machine, operator, conveyor etc. These modules are combined to represent the real system. Each module has its characteristic defined so that it operates similarly to its real-world equivalent. The machine module will have its downtime level, its processing rate, and other variables defined by the person developing the model. These packages have pull-down menus that are easy to use and generally run on a Microsoft Windows background.

Simulation packages have an excellent ability to handle either actual data or statistical data. In this way, the model can operate from real data, such as actual machine downtime per day, or from mathematical formulae that describe the distribution of downtime. The results can be shown in tabular form or on graphs. Very often, it is possible to sufficiently understand the effects of change by just looking at the simulation animation.

Computer models can be sub-divided into two categories: continuous modelling, where the state of the system is recalculated at regular time intervals; and discrete event modelling where the state of the system is recalculated as each event happens. An event may be the arrival of a customer order or the occurrence of machine downtime, for example.

Continuous modelling is generally used to describe continuous steady situations such as the flow of fluid in a pipe. The flow of fluid may increase or decrease but the flow is continuous. The simulation model calculates the state of the system at equal intervals of time.

Continuous modelling may be used to represent environmental systems such as the food-chain relationship between rabbits and foxes. As the number of foxes increases, the number of rabbits decreases. As the rabbits decrease, food becomes scarce and the fox population decreases. The cycle repeats: as the fox population declines the rabbit population increases. The continuous model may provide insight into the relationship between both populations. Changes in state in one factor, over time, affect other factors. Continuous simulation is generally used in the processing industries.

With continuous modelling, the flow being modelled is continuously interacting with the system. Discrete event modelling focuses on the occurrence of events rather than the flow through a system. The state of the system is calculated each time an event occurs rather than at regular time intervals. The discrete event model assumes that the system does not change between the occurrence of events.

Events may be units arriving at a workstation to await processing. This event changes the state of the system. The input inventory buffer increases by one unit. The time at which the unit arrives is noted and can be used to calculate average queuing times. The model searches through the system to determine the time that the next event is due and moves the simulation clock to that time. The next event is activated and the state of the system recalculated. Other types of events include orders arriving, parts being processed, or equipment downtime.

Discrete event models are typically used to simulate systems such as manufacturing, business administration systems, computer networks, or industrial engineering studies. Each of these systems are broken into their component parts and have their characteristics defined.

For example, a clerk who is responsible for responding to customer complaints can process five complaints per hour if uninterrupted. The clerk is also responsible for doing word processing tasks for two other managers. The clerk's level of success in responding to each complaint within 24 hours will depend on the level of complaints received and the level of work supplied by the two managers. Within this system, events would be the receipt of customer complaints, work from managers, and the completion of processing for the complaints and the manager's work. The interval between occurrences is unlikely to be regular and even. In this way the system's state is more related to the occurrence of events than to the passing of time.

The customer complaints would be defined as parts being processed through the system. These parts pass between the system's components. After entering the model they may wait in buffers because the clerk is unavailable. They move to the clerk, are processed and exit the system. The parts can have attributes associated with them that help them to simulate real-life occurrence. In our example, parts or customer complaints may have priorities associated with each one according to its degree of severity. The more serious customer complaints receive attention first. The priority set for each complaint is an attribute associated with that complaint. The model can be designed to look for the attributes of complaints in waiting and choose the most serious for the attention of the clerk.

The model is generally set up to collect data relating to the parts — for example, the number of parts processed or the average waiting time. In our example, this would be the number of complaints processed and the average waiting time. The model is set up to collect the necessary data while the simulation is running.

The clerk coping with customer complaints and the work from managers is a simple example of a real work situation. Simulation enables the scenario to be represented and understood. The effect of changing how complaints are managed can be evaluated, and the results predicted without ever having to set up pilot programs.

SUITABILITY

The suitability of simulation depends on the complexity of the problem and the objectives of the study. Simple systems which have virtually no random elements can be modelled using equations and a spreadsheet. When the aim of the study is to provide only rough estimates on the system's performance, it is often feasible simply to use averages to describe variation. In some situations, when associated costs are low, it is feasible to experiment with the actual system. The changes can be implemented and the system's performance monitored. Simulation is not the cure-all solution for all systems' problems. There are often alternatives that are more suitable to the constraints and requirements that exist.

However, many complex real-world systems cannot be readily experimented with and can only be modelled using computer simulation. Simulation packages are capable of modelling the random variation and interrelated links between various system components.

Simulation also allows the modeller to test the performance of the system under a variety of operating conditions. Experiments can be designed to evaluate the effects of a wide range of alternatives on the performance of the system.

It also acts as insurance against poor decision-making by helping to avoid oversights in planning for change. Computer simulation can help in understanding waiting time, queuing principles and how the addition or withdrawal of resources affects the system. These variables can therefore be planned for at an early stage in the life of the project. Simulation is often the key to success for projects that would otherwise be heading for failure.

In the manufacturing context, simulation enables the manufacturing engineer to estimate the effects of changing the batch size or the production control system, or of introducing group technology. The manufacturing environment typically represents a complex system. Changes to components of the system can have "knock-on" effects on a range of activities. The broad impact of these changes can only be gauged through the use of simulation technology. The following is a list of typical problems which simulation can aid in resolving.

- Designing production lines to:
    ◊ Maximise through-put
    ◊ Minimise manufacturing cycle times
    ◊ Improve capacity
    ◊ Test new production systems
    ◊ Improve delivery reliability
    ◊ Manage maintenance and other resources

◊ Cope with variation in customer demand
◊ Maximise return from investments
◊ Evaluate layout alternatives
◊ Minimise time lost in changeovers
◊ Evaluate optimising conveyor and AGVs (Automated Guided Vehicle systems)
- Designing warehouse systems to:
   ◊ Have effective order picking and delivery scheduling
   ◊ Determine resources needed
   ◊ Allocate shelf space
   ◊ Minimise delays in loading and unloading trucks
   ◊ Calculate quantities of material handling containers needed
- Re-engineering the organisation to:
   ◊ Redesign customer service logistics
   ◊ Redesign customer credit applications
   ◊ Redesign customer order processing system
   ◊ Analyse process flows
   ◊ Contain costs and finances.

Simulation can help managers and manufacturing engineers with a broad range of problems. However, computer simulation has some drawbacks. It can be time-consuming and expensive to develop models. It takes time to collect the data needed for the model, to create the model, to verify its accuracy, to design experiments and evaluate results. Time constraints within the project may make simulation impractical.

Simulation models of systems with random elements cannot provide a single accurate solution. Rather, each run of the model produces an estimate of the real-world system's expected performance under the same conditions. The random nature of the real-world system means that there are no exact answers.

The model should be run several times and the results treated as estimates which describe the system's likely performance. Random variation in the real world means that the model will not exactly duplicate the real world's performance for a given set of conditions.

BASIC CONCEPTS ASSOCIATED WITH DISCRETE EVENT MODELLING SOFTWARE

Today, many software packages are designed to cater for specific industry requirements. This means that packages are produced which are geared for the processing industry while others are focused on the manufacture of discrete units. There are many similarities between the way many of these packages function. In general, discrete event packages for the manufacturing industry provide the user with pre-programmed modules which represent machines, buffers, machine operators, event generators, etc. The programming environment consists of pull-down menus and dialogue boxes. The dialogue boxes and pre-programmed modules minimise the level of programming required. Many of the packages enable the modeller to program in the C language or a customised version of a general purpose language. The pre-programmed modules are represented as objects or icons on the screen. They can be laid out on the screen to represent the current factory layout. Machines, buffers, and people are positioned on the screen to replicate the real world.

In general, modules are customised via dialogue boxes. Each dialogue box provides the modeller with an opportunity to describe the icons' operating characteristics. For a machine these would include mean time to failure, mean time for repair and processing speed. For a buffer, they would include maximum capacity for inventory and initial stock level at the start of the simulation run.

These modules, generally, gather information as the simulation model is running. For example, the buffer will measure the average quantity held during the simulation run and the average queuing time for inventory. The machine will gather information such as its utilisation rate and the quantity of units processed through it.

Once the screen is laid out to represent the real world system, the modeller defines the product routing through various machines. The flow is defined for each part type.

A part generator at the beginning of the process generates parts to be processed, perhaps in a similar manner as that in which customer orders are received. Packages also allow the modeller to use condition-based logic within the model. This enables the modeller to experiment with various ideas such as varying production control techniques.

The parts generated represent parts moving through the production area. The parts can have attributes associated with them such as a priority number that is used by the model to schedule production. Attributes can also gather information relating to the part. As the part passes through buffers and machines, these modules can update the parts attribute. For example, if the cost of the part varies depending on the machine that processes it, then each machine can be set up to change the cost attribute associated with the part. As the part exits the simulation, the model notes the production cost for the part.

The above description provides a general introduction to manufacturing simulation packages. It serves to introduce the concept to readers who may be unfamiliar with simulation. The actual simulation run is driven by an algorithm which is quite complex, however. The main components that make up this algorithm are:

- *System state*: The system state is a collection of variables that define the system at any moment in time;

- *Initialisation routine*: A sub-program that establishes the initial variables to define the system at the start of the simulation run;

- *Simulation clock*: A variable which represents time within the simulation;

- *Event list*: Defines the next event to occur and the time at which it will occur;

- *Timing routine*: Determines what is the next event to occur and when it will happen, and updates the event list with this information;
- *Event routine*: Updates the system state variables to reflect the occurrence of an event;
- *Report generator*: Gathers information from the model during the simulation run;
- *Random generator*: Provides random numbers that are used to generate the occurrence of events randomly;
- *Main program*: Co-ordinates the various sub-routines that combine to form the simulation program.

Figure 6.1 provides a rough overview of how the model operates.

Using Computer Simulation

Computer simulation should be used as a tool to help management and engineers to understand the effects of proposed change. It should not become an end in itself. The computer simulation project has satisfied its goals when it supplies sufficient information to help and direct decision-making.

It is generally impossible to perfectly replicate a real system. The real world is continuously changing, therefore any attempt at 100 per cent computerised replication is futile. The aim of the simulation team is to understand the level of similarity required to satisfy the project goals and to achieve that similarity within the allocated time period.

The steps to modelling are in three stages as follows:
Preliminary work which consists of:

- Developing a team approach
- Defining the problem
- Documenting assumptions
- Collecting the data
- Creating the model.

FIGURE 6.1: OUTLINE OF THE ALGORITHM USED TO CONTROL THE SIMULATION

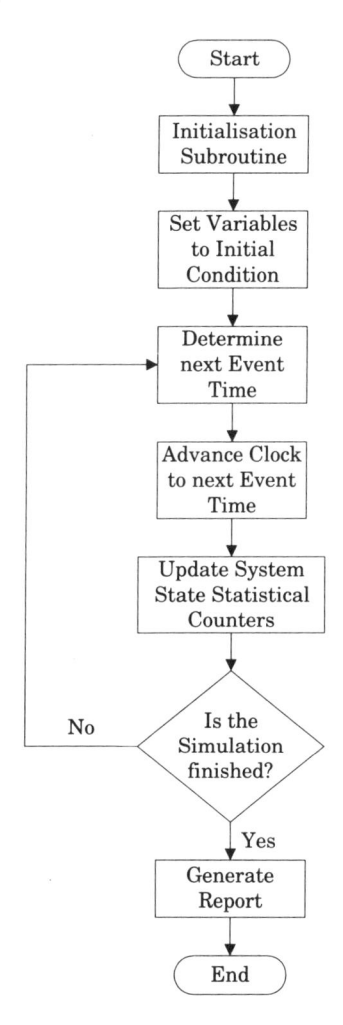

```
Start
  ↓
Initialisation Subroutine
  ↓
Set Variables to Initial Condition
  ↓
Determine next Event Time ←──┐
  ↓                          │
Advance Clock to next Event Time
  ↓                          │
Update System State Statistical Counters
  ↓                          │
Is the Simulation finished? ──No──┘
  │ Yes
  ↓
Generate Report
  ↓
End
```

Analysis work which consists of:

- Verifying, validating and judging model credibility
- Designing experiments
- Analysing the results.

Implementation guidelines which provide some recommendations regarding simulation projects.

Preliminary Work: Developing a Team Approach

It would be wrong to assume the simulation modeller can work on their own and return to the manager with a report describing the solutions to the problem. Complex real-world problems require a team effort to develop useful models that are able to aid decision-making. Some managers consider it feasible to develop a universal model that can be used to solve all problems. This tends not to be practical. The real world is constantly changing, therefore it would be a mammoth task to input all the variables affecting the real-world system into the model and then have to update it as reality changes. Instead, simulation models are designed to help with specific problems. Assumptions are made to reduce the work involved in developing the model and to facilitate possible constraints regarding the availability of data or the capabilities of the software. It is important that the person who commissions the simulation project also partakes in the project in order to gain an understanding of the limitations of the model.

The customer or manager needs to be specific regarding the problem, and the objectives of the study. Assumptions will be made during the project which will need to be approved by the manager. As the customer for the project, this person plays a vital role as the team member who provides direction for the project.

The project team should also have a member who is suitably expert in the real-world system — the system expert. In the manufacturing context, this person may be the industrial engineer or the senior supervisor. Even if the real-world system does not yet exist — for example, if a warehouse is being designed — then there should be a member of the team who is familiar with warehouses in general. This person should be very familiar with the workings of the real-world system and should be able to advise on the level of detail needed within the model to enable the objectives to be reached.

The third member of the team is the modeller — the person who is familiar with the simulation package and with modelling concepts. This person is responsible for creating the model and analysing the results from a statistical perspective.

A fourth member of the team should be from the information services department. This person should be able to assist in data collection for the project. Very often data is available on and can be extracted from the factory's computer system and used as input data for the model. Again, the real-world system expert should check to ensure the accuracy of data supplied by these means.

For the purposes of clarification, the steps to developing a simulation model will be applied to a manufacturing situation. The simple scenario consists of two machines with an inventory buffer between them. The first task for the team is to define the project's objectives.

Defining the Project's Objectives

Specific quantifiable goals help the team to focus on the section of the real-world system that needs to be modelled. Clear goals enable the team to determine the level of detail that will be required in order to attain sufficient accuracy within the model. The following objectives are listed for our example:

- Objective 1: Determine if throughput produced through the machines is significantly reduced if a maximum inventory buffer of 53 units is set between the two machines.

- Objective 2: Determine if the manufacturing cycle time through these two machines significantly reduces due to the maximum inventory level being set.

The objectives should also indicate any time or budget constraints on the project. These constraints will dictate the level of detail that can feasibly be modelled.

All influences on the model are listed and the team decides which aspects of the real world system to model. In our example, the team has to decide if the mechanics of the production scheduling system should be modelled and data generated within the model, or whether the modeller should look at historical records and input this historical data directly into the model. The overall objectives of the study will determine how the system will be represented.

Assumptions

The role of the system expert is crucial in developing appropriate assumptions for the model. Assumptions enable the team to discard the need to model all aspects of the system. It is important, however, to document the assumptions so that those who are evaluating the results of the model understand the assumptions built into it. In this way, any incorrect assumptions are likely to be noticed and highlighted for correction.

The following assumptions were made regarding the model being developed:

- Assumption 1: Weekly or monthly preventative maintenance will not affect the sizing of the buffer between operations. This may be reasonable if both machines are taken down for preventative maintenance work at the same time.

- Assumption 2: Operators go on breaks at the same time and therefore the breaks have a minimal effect on the outcome.

- Assumption 3: Operators are always available to work the machines. The area does not suffer from absenteeism.

- Assumption 4: The quality produced from the first machine is sufficiently good to avoid having to model the reject rate.

The system expert should be able to validate the veracity of each of these assumptions.

Collecting the Data

The team will often be limited by the level of information available. For example, the mean time to failure or the changeover time for equipment may not be available. The information services analyst may supply some data or special forms may have to be generated to collect additional information. The system expert will need to indicate the quantity of data needed to accurately represent the system.

Historical data can be downloaded directly to the model or it can be analysed to identify distribution trends. Using historical data means that the model can only be designed to represent events that have actually happened previously. This means that a large quantity of historical data is often required to reflect the variation within the process. However, when the historical data can be linked to a probability distribution, the distribution can be used to model the random variation. For example, the following frequency distributions are often used to model random occurrences:

- The normal distribution is used for describing the frequency of events due to natural occurrence. For example, the time to manually perform a repetitive task, or variation in units produced.

- The exponential distribution is often used to describe the time between the arrival of customer orders, or the average time between machine breakdowns.

- Uniform distribution is often used to estimate random events when only limited historical data is available. It defines the variable as equally varying between two points, a and b, as illustrated below.

FIGURE 6.2: BASIC SHAPE OF DISTRIBUTIONS

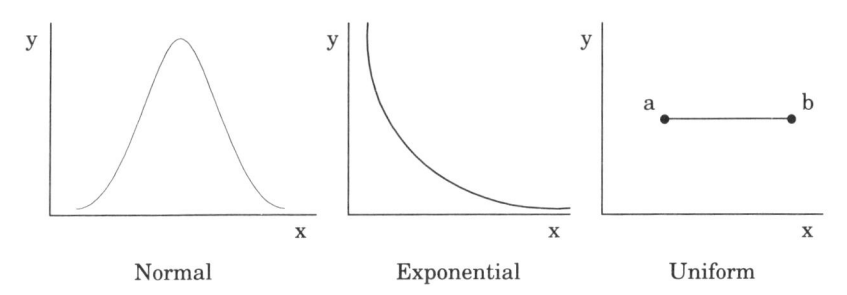

Normal   Exponential   Uniform

These and other distributions can be specified to describe the occurrence and duration of events. Many of the simulation packages offer statistical features that can be used to model random events instead of historical data. These statistical features can evaluate data and determine the suitability of the various distributions.

A useful approach to data collection is to focus on building a rough model initially. This rough model will highlight those sections of the model where time invested in more detailed data collection will directly improve the accuracy of the model.

Small changes can be made with input data for model components. If the output data changes significantly, then the team should ensure accurate data for this variable. For example, when modelling a complex production line, the rough model will generally direct the team to collect detailed information on the bottleneck process because the quality of this data directly affects the accuracy of the model in imitating the line's performance. This approach may reduce the time invested in data collection without reducing the model's accuracy.

Create the Model

The first step in building a model is to sketch it out on paper. Draw out how the machines are positioned in relation to each other. Draw a flow chart describing the product routings. Then lay out the pre-programmed modules such as machines, human resources, buffers, etc., from the software's library of modules. The product routings are defined and the data entered. Simulation packages that are geared for manufacturing generally supply sufficient pre-programmed modules to enable the majority of systems to be modelled. However, when no suitable module is available, the packages often enable the modeller to develop their own modules using the C language or FORTRAN.

In our example, a parts generator, two machine modules, two buffer modules and an exit module were used and laid out as illustrated in Figure 6.2 below.

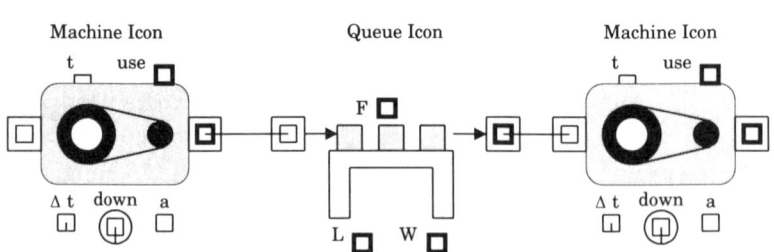

FIGURE 6.2: EXAMPLE OF MODEL LAYOUT*

* Reproduced from the simulation package "Extend" with the kind permission of the company Imagine That!.

The product routing and processing data is entered using dialogue boxes and flow lines. The model is set up to gather information relating to throughput and manufacturing cycle. Both of these metrics are measured as each part exits the model.

Analysis Work: Verifying, Validation and Model Credibility

Verifying the model means checking to ensure that its construction is as planned. It is a general check to ensure that the system that was hand drawn during the creating stage has been converted into a working program. This can be checked by simplifying the input data so that the model's output can be verified by manual calculation. This step ensures that the team has made no mistake in specifying product flows or equipment processing rates. Verification check ensures that the model operates as intended.

Validation of the model evaluates its effectiveness in imitating the real-world system. It tests the output from the model against the output from the actual system to determine the level of correlation. A validation test may consist of loading historical data and comparing the results produced from the model with the results of the real system. Another approach to validating the model is to present simulation results along with real-world results to a team of people who are familiar with the real-world system. The team is asked to segregate the results into those generated by the model and those from the real world. Their ability to differentiate between the real-world results and the simulation results relates to the effectiveness of the model in simulating the physical system. This is known as the Turing test. A valid model means that results attained from experimenting with it are representative of the results that would come from experimenting with the physical system.

Successful validation creates credibility for the model. Credibility means that management believe in the model and consider its results in decision-making. Credibility is essential for effective simulation; otherwise the results from experimentation are ignored. Credibility is also aided by working closely with management and the system expert throughout the project. In this way, their understanding and appreciation of the model improves. Model animation also helps the model to achieve credibility, since management can observe the system in action and see how the results are generated.

Comparing Alternative System Configurations

The power of simulation comes from the ability to test and compare alternative systems without having to implement them. The model is easily changed to describe the new system and re-run to generate performance results.

Since many simulation models contain random elements, the model will generally provide some variation in its results. It is important to run the simulation a number of times before drawing conclusions. Statistics can be applied to analyse the level of variation with results and recommend the number of simulation runs needed to minimise error. Many of the simulation packages offer advanced statistical features which will perform all necessary calculations.

The animation available in modern simulation packages conveys a lot of information regarding the model. The animation will show machine downtime and inventory build-up in the example of the manufacturer investigating inventory control techniques. It will show how resources are being utilised in the initial example of the clerk processing customer complaints. These movie-style presentations can provide powerful evidence of how the process currently operates and how changes can improve it.

In our example, the model of the current production system was run five times and the results noted. The inventory buffer is set to a maximum of 53 units and the simulation run five times to produce five sets of results. These are compared to establish if the maximum limit on the inventory buffer results in a decrease in throughput from the old system, and if the manufacturing cycle time is reduced. The average indicates that the cycle time is reduced by setting up a maximum inventory level for the buffer, and that the throughput is slightly reduced. These findings are presented in the graphs.

The team may want to know the statistical significance of the findings. Is the reduction in cycle time and throughput a chance result considering that the model output is variable or is the reduction statistically significant at a defined confidence level? The following pairwise comparison is performed to check for significance at the 97.5 per cent level. (An introduction to pairwise comparisons is provided in Appendix B.)

FIGURE 6.3: GRAPHICAL COMPARISON OF SIMULATION OUTPUT

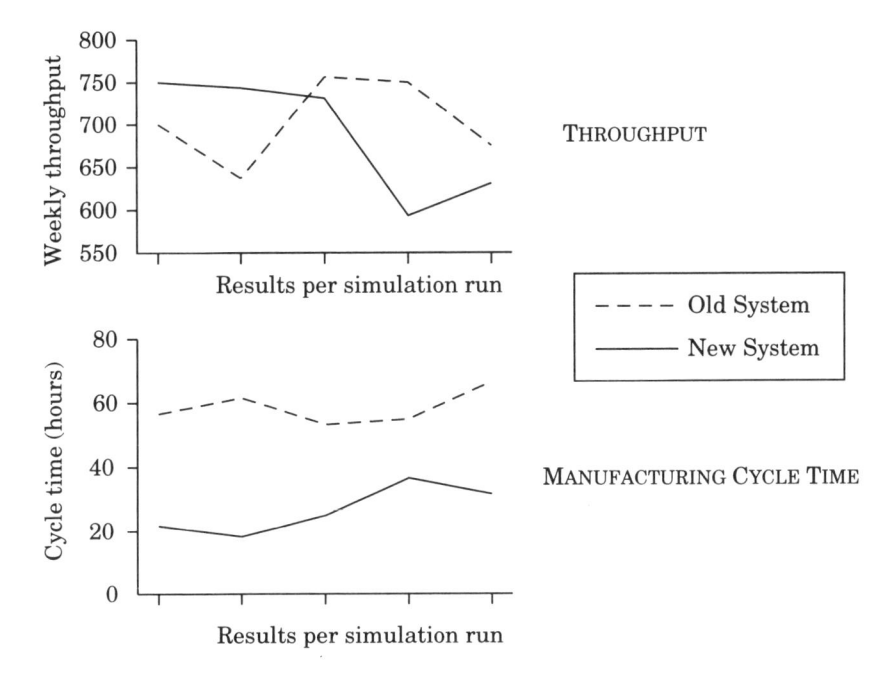

Table 6.1 shows a pairwise analysis of the results.

The pairwise comparison confirms that the cycle time is reduced with 97.5 per cent confidence, but that there is no statistically significant reduction in throughput at the 97.5 per cent level of confidence.

The project achieves its goals by demonstrating that manufacturing cycle time is potentially reduced by up to 52 per cent when the inventory level in the buffer is limited to a set maximum. At the same time, the model predicts that the reduction in throughput may be slight, as indicated in Figure 6.3.

TABLE 6.1(A): PAIRWISE ANALYSIS OF THE SIMULATION OUTPUT: THROUGHPUT

	1	2	3	4	5	Sum	Mean
Old	698	639	761	753	680		
New	748	744	730	595	634		
Difference (X)	–50	–105	31	158	46	80	16
(X–Mean)^2	4356	14641	225	20164	900	40286	

Notes: Testing the Null Hypothesis that the old system throughput is greater than the new system throughput. Estimated population standard deviation = 100.3569. Standard error = 44.88095. The sample mean lies 0.356499 standard errors from zero, the hypothetical mean. The test is to verify that the old throughput is higher, therefore a one-tail t-test is used at 97.5% confidence and 4 degrees of freedom. This t-value is 2.776. The deviation from the mean is less than this number of standard errors (0.356499 < 2.776); therefore the difference is not significant at 97.5% level of confidence. There is no significant evidence to say the old system throughput is higher than the new system throughput.

TABLE 6.1(B): PAIRWISE ANALYSIS OF THE SIMULATION OUTPUT: MANUFACTURING CYCLE TIME

	1	2	3	4	5	Sum	Mean
Old	55	61	52	54	67		
New	22	20	26	36	33		
Difference (X)	34	42	26	18	35	154	31
(X–Mean)^2	8	116	21	167	15	327	

Notes: Testing the Null Hypothesis that the old system manufacturing cycle time is greater than the new system manufacturing cycle time. Estimated population standard deviation = 9.05. Standard error = 4.05. The sample mean lies 7.60 standard errors from zero, the hypothetical mean. The test is to verify that the old cycle time is higher, therefore a one-tail t-test is used at 97.5% confidence and 4 degrees of freedom. This t-value is 2.776. The deviation from the mean is greater than this number of standard errors (7.6 > 2.776); therefore the difference is significant at 97.5% level of confidence. There is significant evidence to say the old system manufacturing cycle time is higher than the new system manufacturing cycle time.

Implementation Guidelines

The choice of simulation package will have a major impact on the feasibility of the simulation. The following features should be considered when purchasing a new package.

User-Friendly and Capable

The software should be easy to use and capable of performing the simulation. Many of the suppliers of simulation software offer demonstration disks that enable the purchaser to experiment and determine its ease of use. Sample simulation, similar to the project being undertaken, may be provided with the demonstration disks. The purchaser should contact the software company directly to discuss the suitability of the software for the project. Important questions include: Can the package cope with the size of the model envisaged? How much time does it take to execute programs? Does the package enable the user to describe condition-based logic that may be needed if, for example, a production scheduling technique were being evaluated?

Animation

The user will need to decide if animation is required for the simulations. Animation tends to make the software more expensive. When animation is important, the purchaser should ensure that the graphics are in bitmap format instead of character graphic. The resolution should be such that the pictures are easily recognisable. Their movement across the screen should ideally be smooth as opposed to jerky. Some of the packages can use CAD drawings as background to the moving images. The overall image portrayed should aid management in interpretation of the model results.

Statistical Features

Statistics is a critical part of simulation. Most of the better packages have statistical features that perform many of the necessary calculations. These packages can fit historical data to common probability distributions. Packages should be able to cope with a range of probability distributions. The statistical features of the package should be able to analyse the output from the model and aid in making statistical inferences.

Customer Support

Customer support should be provided with the package. Customer support may include the availability of training courses and on-line telephone support. Telephone support is usually conditional; the user is expected to continuously purchase upgrades as they become available. The user should enquire as to the cost associated with upgrades and the frequency with which they are released.

Output Reports

Packages should provide standard reports on such parameters as machine utilisation, average inventory levels in buffers, quantities of units processed. Parameters such as these should be collected as standard reports. Output reports should be easily set up to collect other information such as manufacturing cycle time or cost incurred. The model should be able to present the information graphically or, at least, to store the information in files that can be imported into presentation packages.

Pitfalls to Successful Simulation

There are a few common pitfalls associated with attempting to resolve problems through computer simulation. The first pitfall is the failure to have a clearly defined goal for the project. The modeller receives vague instructions regarding what the simulation should achieve. Models tend not to be universal problem-solving tools. Therefore when the problem is initially poorly defined, the simulation model that is built often does not achieve its goal.

Very often, the raw data is not available in order to simulate the real-world system. The production facility may not collect downtime performance or may not provide information regarding times to perform task. In this environment, simulation becomes a major undertaking, as the team has to carefully define the information required and develop systems to gather it.

For many projects, the time required to gather this information makes simulation unattractive.

Part of the simulation project is to determine the level of detail required to model the real-world system. Excessive detail means that effort is wasted in modelling variables that are not significant; on the other hand, insufficient detail results in inconclusive output. The team will have to identify the critical characteristics of the real-world system that need to be modelled and to keep the model simple by focusing only on these variables. Excessively sophisticated models take longer to build, require more computer time to run, and need more maintenance work to keep up-dated. It is recommended that the team should start out with a simple model and build up the level of detail within it as required by the project's objectives.

It is important for the project to maintain good communication with its sponsor or customer. The team needs to verify that assumptions made during the development of the project are acceptable. Models can give misleading impressions unless the people reviewing the results are aware of all of the limitations built into them.

Conclusion

Simulation is a powerful technique for demonstrating the benefits to be derived from using alternative manufacturing techniques without having to experiment with the physical system. It is a less expensive approach to experimentation. It involves constructing a replica of the real-world system that can be used to represent real-world experiences within the factory. The graphics used in modern simulation software make the experience similar to watching a movie. What if scenarios can be evaluated and accounted for in designing the new systems? There are so many variables affecting change programmes such as a JIT programme or a Process Reengineering programme, that often computer simulation is the only tool available to avoid costly miscalculations. In effect, computer simulation allows the innovator to experiment with virtual processes before attempting implementation.

However, computer simulation has its limitations. It is important to understand these limitations in order to use the technology correctly. Clear goals are needed for the simulation project. The project should be a team endeavour consisting of the relevant manager, the real world system expert, an information services analyst and the simulation modeller. The output from models of systems containing random elements should be considered as estimates of how the real system would perform. These estimates should be analysed statistically before drawing any conclusions.

Computer simulation is set to become a very important management tool, especially as the software packages become more user-friendly and inexpensive. They enable management to experiment and adapt new ideas that will make their factory more efficient.

Chapter 7

PROCESS REENGINEERING

INTRODUCTION

In order to understand the concept of process reengineering, it is necessary to examine the origins of industry. Adam Smith (1776) is credited with developing the concept of the modern factory. Before this time, goods were produced by craftspeople who worked alone producing a complete product. The cobbler was a tradesman producing shoes, the tinsmith producing pots and pans. Smith described how a pin factory managed to produce thousands of pins each day by dividing up the labour involved in pin-making amongst the employees. Instead of one person producing a complete pin, each person performed one or two sub-tasks, the combination of which resulted in pins being produced. According to Smith, a job like pin-making can be divided into sub-tasks such as drawing the wire, straightening, pointing, grinding. Each person is given responsibility for performing one of these tasks. Because each person has only a sub-task to perform, they become specialised and proficient at that task, thereby improving the overall efficiency in making pins. The sequence of tasks could be considered as the process for making pins.

Processes within manufacturing are generally easily identifiable and are well understood by manufacturing personnel. However, there are also processes operating within administrative functions. Alfred Sloan applied Smith's concepts to the administrative functions in General Motors. Jobs such as receiving customer orders and purchasing raw materials were broken down into their component tasks and people were given responsibility for performing each task. Processes were set up for financial budgeting, payroll, order entry, production scheduling, and other administration functions.

Smith's concepts were a great improvement on the old system of craftspeople. Virtually every department breaks down its main job into sub-tasks which are distributed to various people to perform. As each person becomes proficient in the task allocated to them, the efficiency of the process improves.

Many of the processes in use today were originally developed many years ago, before the advent of computer technology. Computer technology has increased the design options available when developing processes. It now means that many of the tasks, such as performing manual calculations or generating reports, can be automated. This new technology enables people to take on extra responsibilities and tasks, resulting in a less fragmented process.

Computers are changing the way we do business. Many of today's processes are out of date because they were designed in an era when priorities were different. In times when a product's life extended over several years, the time for new product introduction was less important. Today, with product life cycles as short as three or six months, the cycle time for new product introduction is critical. Similarly, as we move away from mass markets, companies will need to alter their processes to supply customised product. Many companies are trying to cater for these changes by adapting existing systems, when what is actually needed is a complete process redesign.

Process reengineering is about redesigning processes that have become inefficient and ineffective in light of recent technological advances. Effectiveness means that the process achieves its aims, while efficiency means that it consumes the least amount of resources in so doing. Because of the archaic nature of some processes, there is often a lot of scope for improvement through reengineering. Projects can yield improvement gains of 50 and 100 per cent.

Process reengineering is an improvement tool for management because they are the process owners. They are most aware of the limitations of current processes and have the authority to instigate change. They can establish a project team empowered to reinvent the way the company achieves its aims.

The powerful and radical nature of process reengineering may make it attractive to companies facing bankruptcy or companies who wish to shake up their market segments; they may be in need of an urgent and effective overhaul. The implementation times associated with reengineering projects tend to be short compared to the level of improvement gained.

Companies may perceive process reengineering as a means of cutting costs and gaining market share on competitors. In the 1990s, the price of personal computers was slashed several times. Those companies that were not prepared for the price cutting inevitably lost market share.

This chapter describes process reengineering from a manufacturing perspective. It describes techniques for developing new processes and evaluating existing ones. An overview of modern computer technologies is presented which enables repetitive tasks to be automated. Application of these techniques produces more focused and efficient processes which improve the factory's overall performance.

DEFINITION OF PROCESS REENGINEERING

Process reengineering was defined by the originators of the concept, Hammer and Champy (1994), as:

> abandoning long-established procedures and looking afresh at the work required to create a company's product or service and deliver value to the customer. It means asking this question: "If I were re-creating this company today, given what I know and given current technology, what would it look like?"

Hammer and Champy describe a reinvention of the company's existing processes. They coined the term process reengineering to describe the systematic approach taken to overhaul the company's systems.

SUITABILITY OF PROCESS REENGINEERING

Process reengineering should be a useful improvement technique for any manager of a process. The plant manager can adopt process reengineering to improve delivery reliability or to reorganise staff, the finance manager can use process reengineering to improve the purchasing system or the budgeting process.

Processes which would benefit from reengineering are easy to recognise; they can generally be described as containing high levels of non-value-added work such as rework or transposition of information from one format to another. Other features include slow processing times and erratic performance. The following is a description of how a factory supplies a commit date for orders to its sales office.

The sales office is located many miles from the factory. Sales orders arrive at the factory via the computer network. The factory has to respond with a commit date for delivery of each order. The commit date depends on the level of inventory and factory capacity available.

The inventory status is reported via the internal factory software, which is not compatible with the software that delivers the customer order. The factory capacity available is monitored using a manual system. The scheduler needs to access each of the three sources of information in order to give a commit date for each order. Each of the sources has to be updated independently, resulting in the same information being manually transformed.

This example shows how responding to the sales order with a commit date involves exchanging data from a variety of sources which are mutually incompatible. Improving any of the individual systems will not provide much overall improvement. Significant improvement is only achieved when the administration process involved is overhauled through process reengineering. The description also suggests that computers could be better used to automate this task and perhaps other tasks within the process.

Another example shows how a company arranges for the ordering of maintenance spares.

The existing method for ordering spares evolved when the factory was small, operating on only one shift. Now, three shifts operate, but the approach to spares reordering has remained the same. All spares were ordered during the day on an informal basis by technicians. Technicians monitor their spares supplies and replenish them as they consider necessary. Spares are stored in a variety of places including individual toolboxes that are locked when the technician is not on the shift. This results in delays in finding spares. The system often results in technicians being telephoned during their off-shift to find the whereabouts of spares.

The technicians decide which spares to purchase from their experience of what has historically broken on the machines. When a new fault occurred, there may be a two- to three-day delay before the spare arrives. Sometimes the spare is not ordered in time and equipment remains down pending its arrival. In some cases, multiple vendors are used to supply similar parts, leading to excess part numbers and unnecessary variation in spare parts. Much technician time is wasted counting and reordering spares.

This description of ordering spares indicates that spares availability is generally erratic. The approach used is very *ad hoc* and poorly defined. It could be concluded that there is no effective process for reordering spares.

Both examples describe processes that would benefit from redesign. The existing systems are both inefficient and ineffective. They offer excellent scope for major improvement through reengineering.

APPLYING PROCESS REENGINEERING

Process reengineering is described in three steps. The first step, *preliminary work*, describes:

- How a process can be defined through the use of process mapping;
- What should be considered as appropriate targets for the team.

The second step, analysis work, provides:

- An overview of modern computer technology;
- Methods of developing alternative process designs.

The third step, implementation guidelines, expands on the typical requirements that new process designs need before being established.

The steps are provided as a guideline for a reengineering effort. They help to demonstrate the steps involved in designing a process. The ideas presented can generally be adapted to the individual needs of any reengineering project.

Preliminary Work

Process Mapping

The first step in process reengineering is to clarify how the existing process operates, how it succeeds in converting inputs into outputs. Process mapping or flow-charting is the technique used to describe how a process operates. Process mapping uses the following symbols and layout to describe the sub-tasks and conditional logic inherent in the process.

FIGURE 7.1: PROCESS FLOW CHART

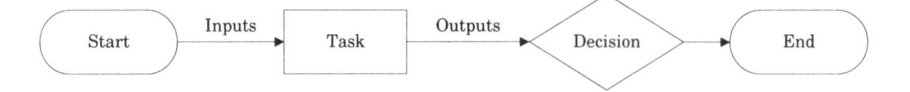

Start → Inputs → Task → Outputs → Decision → End

Few companies have their processes defined in such a format. Therefore, the team will have to convert their general understanding of how the process currently operates into the flow chart format. Figure 7.2 illustrates the production scheduling process described earlier.

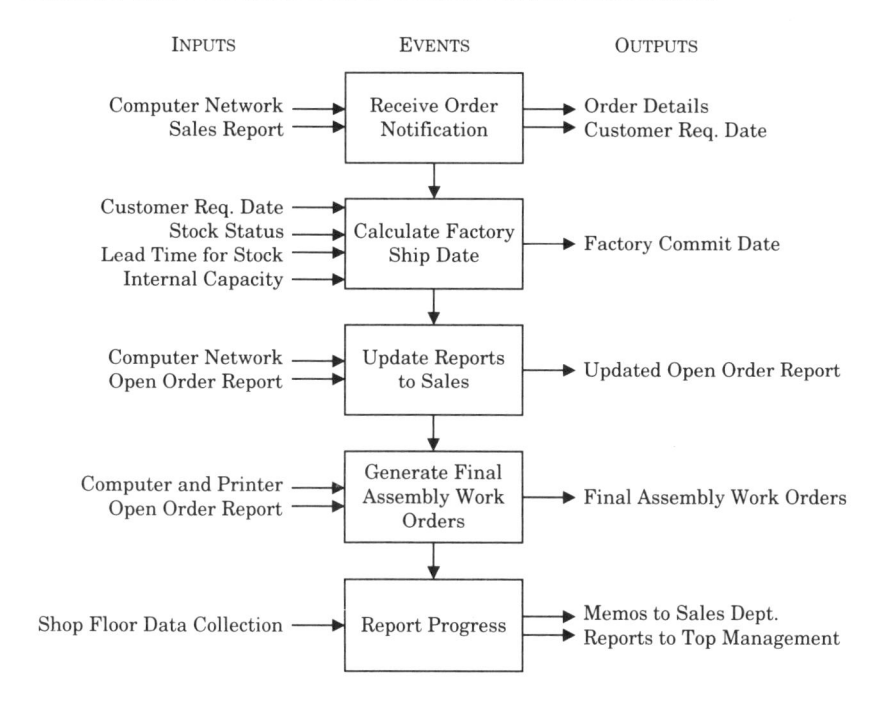

Figure 7.2: Process Flow Chart for Scheduling

Inputs	Events	Outputs
Computer Network, Sales Report	Receive Order Notification	Order Details, Customer Req. Date
Customer Req. Date, Stock Status, Lead Time for Stock, Internal Capacity	Calculate Factory Ship Date	Factory Commit Date
Computer Network, Open Order Report	Update Reports to Sales	Updated Open Order Report
Computer and Printer, Open Order Report	Generate Final Assembly Work Orders	Final Assembly Work Orders
Shop Floor Data Collection	Report Progress	Memos to Sales Dept., Reports to Top Management

Process mapping highlights how the various steps within the process interact to produce the end units. It establishes current practises and provides an initial platform for idea generation. Process mapping also highlights inefficiencies within the process and provides an insight as to how improvements may be made.

Setting Performance Targets

Setting new performance targets means imagining how the perfect process would perform. The team should envisage how they would like to see their company operate in 3, 5, or 10 years time. Idealistic targets may be 100 per cent good product, 25 per cent reduction in costs, 50 per cent reduction in manufacturing lead time, 100 per cent delivery reliability. For the process of controlling spares, it may be that spares usage is reduced by 50 per cent and that there is 100 per cent availability of essential spares. In the case of providing a promise date for a customer order, it may be that the promise date can be provided at the same time as the customer query is received. The new process should be designed with the aim of achieving these targets.

The team should consider comparing their current process and its targets against other competing or similar companies. This exercise may provide the team with new ideas and targets that are proven to be achievable. Many manufacturing problems are common to all industries and therefore there is ample opportunity to discover how other people resolve similar problems. The team researches a broad spectrum of industry and copies the best methods discovered.

Any limitations or constraints on the reengineering project should be clarified at the outset. For example, there will be cost limitations on achieving the goals or the team may be restricted to using existing information technology. The people implementing the changes may also be limited by their span of authority to influence other processes within the organisation. When these limitations are defined at the beginning of the project, much effort and unrealistic planning is avoided.

Analysis Work

New Opportunities from Computer Technologies

Computer technologies have become an integral part of business life. Computers are helping the manufacturing industry to develop new products, reduce costs, reduce lead times and improve quality. The capabilities of computer technologies have a major role to play in process redesign aimed at maximising process efficiency.

Computers are helping us to achieve higher levels of productivity by gathering and analysing data. They are making radical process redesign possible by providing the necessary number crunching, information distribution, and decision support systems.

The range of computer technologies available is almost matched by the range of acronyms available to describe them. Common acronyms associated with computers are CAD/CAE, CAM, CAPP, SFDC, and CIM. These are briefly explained as follows:

- *Computer Aided Design and Computer Aided Engineering (CAD/CAE)*: In its simplest form, CAD/CAE replaces the traditional draughting board of designers with an electronic one. More sophisticated systems enable the designer to generate a three-dimensional model of a part and perform detailed engineering analysis. The environment in which the part will operate can be simulated and the induced stresses on the part analysed. Systems with kinematic capabilities enable the mating characteristics and vibration in use to be checked.

- *Computer Aided Manufacture (CAM)*: Ever since the 1940s, machine development has been trying to replicate human tasks. CAM is the enabling technology for achieving much of this aim. Computer controlled machines are in almost every industry. Some CAD/CAE systems are able to automatically generate the programs required by the production equipment to produce the part. Metal cutting machines, lathes, milling machines, etc. can be programmed directly from the designer CAD/CAE system.

- *Computer Aided Process Planning (CAPP)*: CAPP is very useful for companies producing a wide variety of parts or for a cell system which produces families of parts. A CAPP system analyses the part to be produced, determines which family of parts it belongs to and schedules it with the corresponding production cell.

- *Shop Floor Data Collection (SFDC)*: SFDC involves monitoring activities at each workstation and the progress being made as lots are processed. SFDC provides an accurate record of shop floor and inventory status. It records product history information, which is important for traceability. Information such as the lot identification number, the scrap level generated while processing the lot, etc. is entered into the SFDC system as each lot moves onto the next operation. This information can be compared against the production schedule and any deviation from the plan can be rectified.

- *Computer Integrated Manufacture (CIM)*: Each of the previous computer technologies described involves the automation of a business function. SFDC automates data gathering and analysis, CAPP automates part of the production scheduling task, CAM describes computer control equipment and CAD/CAE supports product design. Computer Integrated Manufacturing links each of these "islands of automation" so that each system can communicate with the others. In this way, information flows freely and can be manipulated by each system. One definition of CIM describes it as:

> the concept of a totally automated factory in which all manufacturing processes are integrated and controlled by a CAD/CAM system. CIM enables production planners and schedules, shop floor foremen and accountants to use the same database as product designers and engineers. (Mortimer, 1985)

The philosophy of CIM integrates islands of computer technology so that they complement each other. All the previous acronyms are subsystems which, when integrated, create a CIM environment. The integration of the parts creates a system that is more powerful than each of the parts operating independently. Companies that purchase technology with a long-term perspective of incorporating it into the overall information technology system will reap the rewards of a CIM environment at minimum expense.

Business databases are also available which enable the vision of the paperless office to be achieved. These business databases perform a variety of tasks such as accepting customer orders, updating the material requirements plan, generating purchase orders for raw materials, paying suppliers, controlling the pay roll, keeping records of employees, managing the financial accounts, generating reports, monitoring inventory, to name but a few. There are a variety of software houses offering packages that can be customised to cater for specific companies' information needs.

The process reengineering team must be aware of the technologies available. More than ever before, they can envisage how the process should ideally operate, because the software is available to enable the idealised process to be established. Process reengineering is about making major improvements in performance that positions the company competitively; this can only be achieved by understanding the capabilities of computers and maximising the process with their help.

Searching for Alternatives

Searching for alternative process configurations can be a daunting prospect. People's jobs will be affected, organisational responsibilities will change. Implementing the new process will mean disrupting current systems and the work environment. These concerns and worries should be ignored as part of the reengineering effort, in the initial creative stages. Solutions to these problems will have to be found in due course, but will only take from creative thinking at this early stage. Reengineering is an innovative process, and worrying about these factors will only stifle creativity. The team must ignore the anticipated difficulties with implementation and concentrate on designing the best possible process.

There are a number of ways the team can approach this, two of which are presented here. The first involves aggressive questioning of the current process to seek improved alternatives. The team can use Kipling's serving-men:

> I keep six honest serving-men (they taught me all I knew);
> Their names are What and Why and When and How and Where and Who.

The overall process and each of its component tasks are systematically questioned in order to seek out alternatives. For example, the "calculate" step in the order acceptance process is further broken down into its component parts and each part is questioned for its suitability.

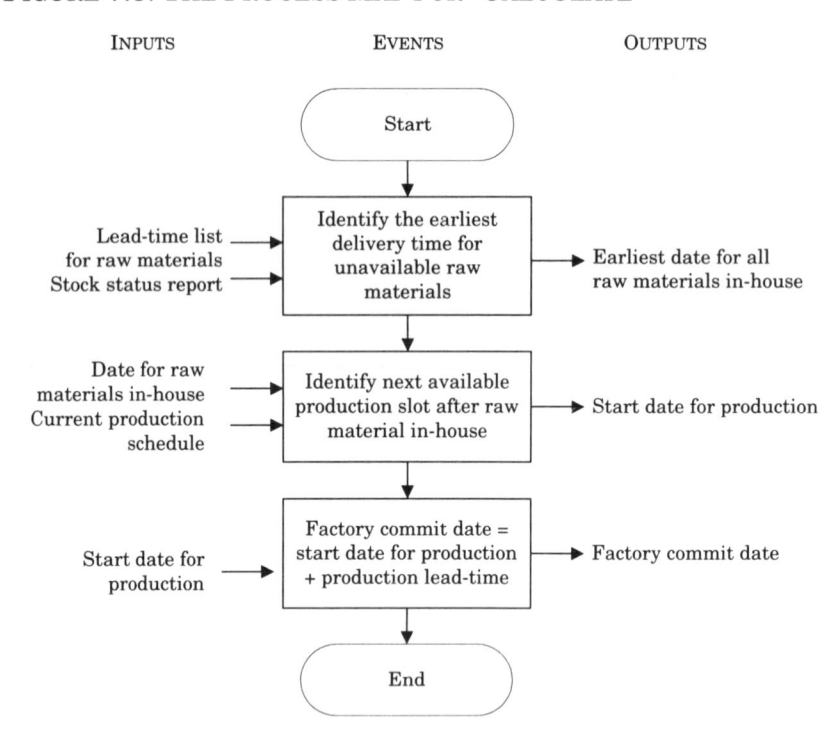

FIGURE 7.3: THE PROCESS MAP FOR "CALCULATE"

INPUTS	EVENTS	OUTPUTS
	Start	
Lead-time list for raw materials / Stock status report	Identify the earliest delivery time for unavailable raw materials	Earliest date for all raw materials in-house
Date for raw materials in-house / Current production schedule	Identify next available production slot after raw material in-house	Start date for production
Start date for production	Factory commit date = start date for production + production lead-time	Factory commit date
	End	

Each of the other steps in the process is evaluated for appropriateness and, if necessary, alternatives sought. Table 7.1 can be used to evaluate several steps in the process.

Aggressive questioning of each of the process steps enables a new process to be designed that is more efficient in the light of new technological and economic changes.

The second approach relates to establishing a new process. When there is no existing process, the team begins with a blank sheet of paper. They can invent the process without any constraints. The team can define the ideal state and work backwards to describe the process that achieves this state. For example, in the earlier description of the spares reordering, it is evident that the method was haphazard. There was no procedure for ensuring that spares existed. The current reordering

Process Reengineering 251

TABLE 7.1: PROCESS QUESTIONING

System Title: Production Scheduling and Control

Function	Why is function performed?	Where is function performed?	Who performs function?	What alternatives exist for each function?
Receive order	To interface the sales and production departments	At factory, located in administration offices	Factory administrators	No alternative
Calculate	Determine factory ship date	As above	As above	This task is performed by specific rules; could be automated and performed by the sales department. The advantage would be speed in communicating with the customer
Update progress reports for sales	Customer feedback	As above	As above	Networking/connecting the sales department to the factory IT system would enable the sales to enquire directly as to the progress, avoiding the need for reports.
Generate work orders	Initiate work	As above	As above	Generating work order depends on inventory status on raw materials. Therefore positioning administrators closer to stores should improve efficiency.
Conclusion				Investigate reorganisation of the work as described above (e.g. sales calculating the commit date and moving administrators to the stores area would enable administrators to take on other tasks such as covering a wider range of product lines.

process depended on the technician remembering that particular spare parts were in short supply. In the absence of a recognised process, the reengineering team may have more freedom to define the ideal process. The starting point is to define the objective for the process — in this situation, 100 per cent essential spares to be readily available at all times.

The team can work backwards from this goal to define a process capable of achieving it. For example, the team may decide that a common storage area is required for all spares. This point raises a series of other issues:

- Who will control the common area on a day-to-day basis and departmentally?
- How will the spares be issued out?
- How will the reorder system operate?

Each of these questions gives rise to another sub-set of questions. As the questions are answered, the shape of the new process takes form. The steps are defined and responsibilities distributed. The flow charts in Figure 7.4 describe how the haphazard approach to reordering spares may be revolutionised to improve availability of spares.

The system described is typical of systems in many factories and is therefore not particularly unusual. This project was a major undertaking for the factory and the improved availability of spare parts dramatically improved equipment uptime levels. It serves to show how a process is designed where none existed before.

Reengineering a process carries many risks because it means changing job responsibilities and descriptions and typically investing in computer technologies. The changes introduced often cross departmental boundaries, as in the case of the spare parts reordering process, which can have implications that need to be planned for. Such radical changes, either to existing processes or in the implementation of a new process, require the support of the factory's senior management.

Process Reengineering 253

FIGURE 7.4: NEW PROCESS MAPS FOR CONTROLLING SPARES INVENTORY SHOWING RESPONSIBILITY

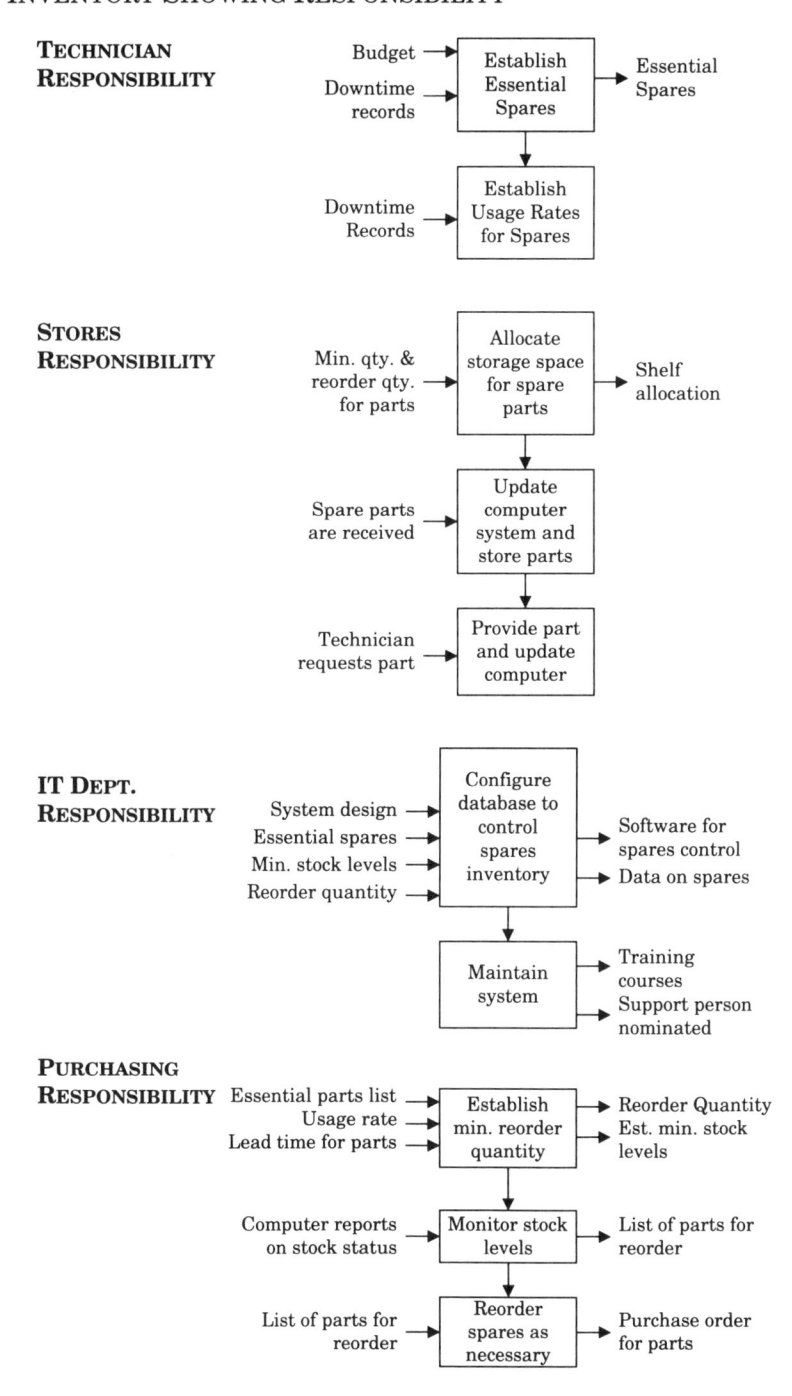

TECHNICIAN RESPONSIBILITY

Budget, Downtime records → Establish Essential Spares → Essential Spares

Downtime Records → Establish Usage Rates for Spares

STORES RESPONSIBILITY

Min. qty. & reorder qty. for parts → Allocate storage space for spare parts → Shelf allocation

Spare parts are received → Update computer system and store parts

Technician requests part → Provide part and update computer

IT DEPT. RESPONSIBILITY

System design, Essential spares, Min. stock levels, Reorder quantity → Configure database to control spares inventory → Software for spares control, Data on spares

→ Maintain system → Training courses, Support person nominated

PURCHASING RESPONSIBILITY

Essential parts list, Usage rate, Lead time for parts → Establish min. reorder quantity → Reorder Quantity, Est. min. stock levels

Computer reports on stock status → Monitor stock levels → List of parts for reorder

List of parts for reorder → Reorder spares as necessary → Purchase order for parts

IMPLEMENTING THE REENGINEERED PROCESS

Implementing reengineering is an exercise in project management, which is covered in more detail in the next chapter. However, before beginning the implementation process, the team will have to define the requirements of the new process. In our example of the spare parts reordering system, new requirements will include:

- A common storage area to be allocated to spares
- Personnel to be hired to manage the spares store
- Job descriptions to be updated for technicians and purchasing personnel
- The information services department to develop and support software that is able to support the spares reordering process. The software may need to be identified and configured to suit the new process
- The finance department to allocate money to the project and to perform a cost/benefit analysis
- Company procedures to be updated to reflect the new process

The team will effectively have to "sell" the entire concept to the various department managers and the factory manager. It is this cross-functional element of reengineering that makes it imperative that the project is initially driven by senior management. Support at this level will be needed to ensure the project receives sufficient attention within the various departments.

Investing in computer technology can be a difficult undertaking. Computer technology is a powerful tool when used diligently, but the team should be wary of using the technology to control complex processes. Processes should be simplified as much as possible in order to reduce the level of complexity. Simple processes tend to be less expensive to automate. Where possible, new technologies should be introduced on a pilot basis before attempting full implementation. Pilot introduction of new technologies helps to generate acceptance for the new technology and to train staff before further investment.

Having identified the new process requirements and received the support of senior management, the team can use standard project management techniques to implement the process. Because implementation of reengineering tends to affect a wide range of people, it can be disruptive to the normal workings of the company. For this reason, reengineering is not like other typical continuous improvement in that it tends to be a one-off project. Continuous reengineering of the spares reordering process, for example, would only introduce confusion and anarchy.

CONCLUSION

Process reengineering offers significant scope for major improvements in company performance. Many company processes were designed before the advent of computers; nowadays competitive advantage can be gained by recognising the potential role of these new technologies. As computer technologies become less expensive in the coming years, their importance becomes critical to remaining competitive.

Process reengineering means recognising those processes within the company that are ineffective and inefficient. The reengineering team systematically maps the existing process, redesigns it to be more effective and implements a new process that favourably positions the company for the next five to ten years. Process reengineering results in gains in the order of 50 and 100 per cent. These improvements are generally only possible through the radical solutions identified with reengineering.

Process reengineering plays a potentially major role in enabling Western manufacturing to survive the manufacturing excellence demonstrated by Asian competitors. As Paul Alaire, CEO of Xerox, who revolutionised his company's performance, said:

> We're never going to out-discipline the Japanese on quality.
> To win, we need to find ways to capture the creative and

innovative spirit of the American worker. That's the real organisational challenge.

It is this pioneering spirit that characterised Western manufacturing at the turn of the century, this same pioneering spirit that is needed to return Western manufacturing to a position of strength. This can be achieved through innovative process reengineering.

Chapter 8

AUTOMATION TECHNOLOGIES

In today's global economy, manufacturing companies are in head-to-head competition with rival firms from different parts of the world. Powerful new economies are emerging from which rapidly growing companies are aggressively winning market share from their western rivals. They have the advantage of a plentiful supply of low-cost labour, guaranteeing low product cost. In order to stay competitive, many western multinationals have moved their production factories to developing countries such as China, India and Thailand. If western manufacturing is to compete, it must become innovative and turn to automation technologies to achieve comparable productivity levels. Manufacturing management must recognise the opportunities provided by automation and develop an investment strategy that ensures the company survives in today's competitive world.

The reason developing countries have been able to adapt so quickly is that much of the technology used today is standardised and easily transferable between countries. In general, many companies use standard technologies to produce goods for the mass market. Most competitors use similar production equipment, have access to the same technology and to semi-skilled personnel. In the contract computer board assembly business, for example, companies like Siemens might sell an off-the-shelf production line capable of producing a specified range of computer boards, Philips might supply the conveyors linking machines, while processing technology such as solder paste is supplied by various raw materials vendors. As a result, the production facility competes for business against competitors who are using similar equipment and raw materials. In this environment, low-cost labour is the competitive advantage that enables developing economies to emulate the success of neighbouring countries like the Republic of Taiwan, South Korea, and Singapore, among others.

From a manufacturing perspective, it can be very difficult to create manufacturing advantage without making some major innovative strides. Western industry is forced to compete against companies using low-cost labour by developing industrial know-how and using modern technologies to achieve flexibility and increases in productivity. These manufacturing technologies are more difficult to transfer across international boundaries because of the high skill levels required to utilise them. Competitive advantage can be regained by adopting advanced technologies such as robotics, flexible manufacturing systems or CAD/CAM which increase manufacturing productivity and flexibility. These technologies enable manufacturers to respond rapidly to changes in the market place and provide a diversified product range to their discerning customers.

There are a number of potential benefits to be derived from choosing and implementing suitable advanced manufacturing technologies:

- *Increased labour productivity*: automation technologies perform repetitive, programmable tasks. Operators are better utilised supervising these new machines. The output per operator increases as process steps are automated.

- *Improved product quality*: Automated processes tend to be more consistent in quality than manual processes. Machines can be continuously improved and reprogrammed to avoid defect generation. In fact, automation of the process is the only means of achieving the high tolerances required by some product specifications.

- *Reduced manufacturing lead-time*: Computer technologies enable the manufacturer to reduce the time from receipt of order to delivery of goods. Information regarding the customer order, the product specification and the relevant machine programmes are downloaded to machines from a central computer system through local area networks.

Automatic changeovers between products means that set-up time is practically zero. The task of co-ordinating and scheduling production is supported by software packages like computer-aided process planning and the use of flexible manufacturing systems. Each of these technologies reduce the time to fulfil customer orders.

- *Reduction of in-process inventories*: The high cost of inventory makes it essential for factories to maximise their use of raw materials. This means minimising scrap generation and moving raw materials from the "goods in" department through the manufacturing process and on to the customer in the shortest possible time span. Data management systems track the whereabouts of materials and the causes of scrap generation. This strategy minimises inventory costs and working capital requirements.

- *Safety*: Automation is often the only available solution to problems related to process safety. One common application of robots is the spray painting of cars. This work environment is not suitable for manual labour; automation technologies offer a practical alternative.

- *Labour shortages*: In many western countries, there is a shortage of people willing to work in a manufacturing environment. These developed economies offer a range of alternative sources of employment in the service sector, and for some companies, automation technologies is a means of expansion when labour is not available.

Many of the modern technologies being implemented today were developed more than 25 years ago. The first Flexible Manufacturing System (FMS), discussed below, was implemented over 25 years ago and there were more than 200 FMSs in operation in the world in 1984. The high cost associated with these initial implementations meant that only large multinationals can afford the investment in technology. While FMSs are still very expensive, much automation technology has reduced in price, reflecting the reduced cost of computer technology and recent technical advances. The following is an overview of a range of automation techniques and technologies available. It includes:

- Machine vision systems
- Automated material handling systems
- Robots
- Flexible manufacturing systems
- Design for automated assembly
- Computer integrated manufacturing.

MACHINE VISION SYSTEMS

Machine vision systems enable automated inspection of parts being produced on the line. In recent years they have begun to replace manual inspection, especially in the component and assembly industries. Automated vision systems have many advantages over manual inspection; with the advances made in these systems in the past 10 years, the investment costs have been drastically reduced; efficiency has also increased. Manual inspection tends to have a relatively low success rate, typically only 80 per cent effective in identifying rejects. By comparison, automated vision systems are very consistent and capable of rapid measurement of product characteristics against defined tolerances. Vision systems are used for purposes other than quality inspection. They can be used to control the routing of products by reading labels attached to goods and sending the material to a prescribed destination. They are frequently used on automated electronic assembly equipment to identify the relative position of work pieces within jigs and to aid in the placement of components. As people become more familiar with the technology, they find new uses for it and incorporate it into different types of machines.

Vision systems operate by performing three functions: image acquisition; processing of the image; and interpretation of the data for the application on hand. Image acquisition uses a localised lighting system and a camera to capture the image. The light source used is generally a series of red light-emitting diodes; the camera is fitted with a filter to avoid stray lighting from the environment distorting the image. Typically, solid-state cameras are used to acquire the image, and an analog-to-digital converter translates the image into a digital format.

There are two common approaches to processing images: binary vision and grey-scale format. In binary vision, the image that is acquired is broken down into pixels, which are interpreted as either black or white. A threshold of intensity can be set; light intensity above this threshold is considered as white while below this level is considered black. Knowing the size of pixels, the system can recognise patterns or calculate the dimensions of the component. The accuracy of the system depends on the number of pixels within the picture frame.

FIGURE 8.1: BINARY VALUE SYSTEM VERSUS GREY-SCALE VALUE SYSTEM

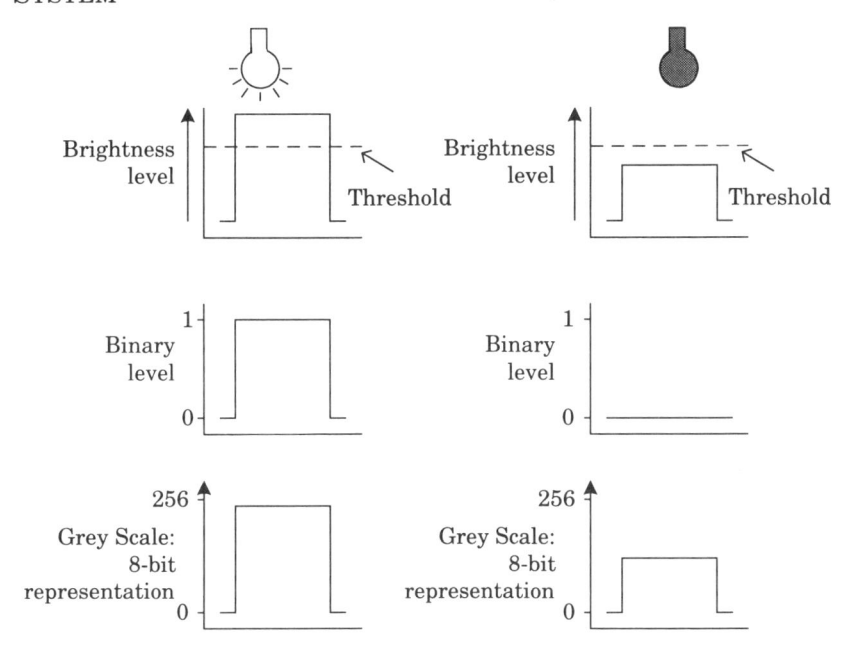

With the grey-scale system, each pixel intensity is classified against a scale using 8 bits of memory, thereby enabling the system to recognise up to 256 variations in light intensity. The grey-scale system enables the surface texture to be recognised

as well as a range of colours. The basic components of a machine vision system are shown in Figure 8.2 below.

FIGURE 8.2: BASIC COMPONENTS OF MACHINE VISION SYSTEM

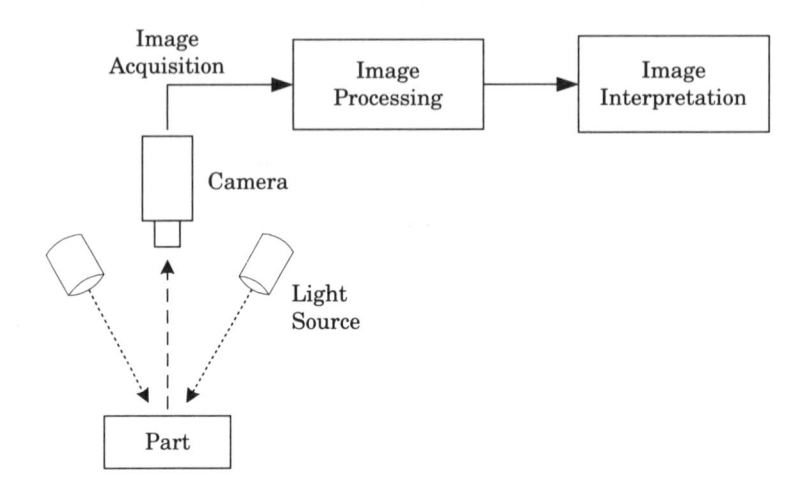

```
     Image
  Acquisition  ──────▶   Image      ──────▶    Image
                      Processing             Interpretation

            Camera

      ◣        ◢   Light
       ╲      ╱    Source
        ╲    ╱
         ▼  ▼
         Part
```

The computer interprets the image by comparing it against an algorithm that defines acceptable and unacceptable product characteristics. The algorithm can be programmed directly into the system, or the system may allow the user to present an acceptable unit whose image is stored in memory, and against which other units are compared. The parts are equated to either the algorithm specification or the image stored in memory; depending on the level of correlation, the part is either passed or rejected.

AUTOMATED MATERIAL HANDLING SYSTEMS

Automated material handling systems are categorised into two groups: *conveyor systems* and *automated guided vehicles*.

Conveyor Systems

Conveyor systems are used where quantities need to be transported between specific locations over a fixed path at frequent intervals. Conveyor systems are generally classified according to the hardware used to transport goods. Typical classifications are: roller; belt; chain; and overhead.

FIGURE 8.3: CONVEYOR SYSTEMS

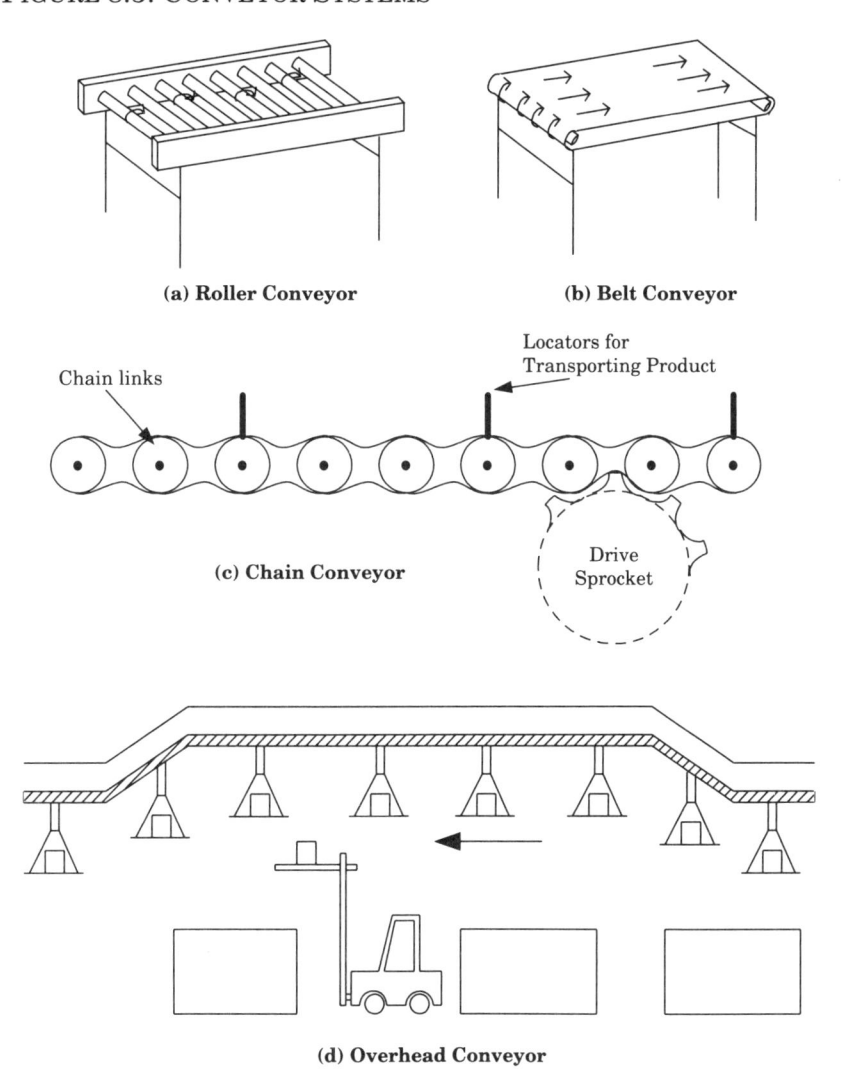

(a) Roller Conveyor

(b) Belt Conveyor

(c) Chain Conveyor

(d) Overhead Conveyor

Roller

The roller conveyor (Figure 8.3(a)) is the most common type of conveyor system. It is relatively inexpensive, and can be either power- or gravity-driven. Roller conveyors are capable of carrying heavy loads and require relatively simple control technology. Constant drive conveyors have their rollers continuously rotating, gravity conveyors have an adjustable slot to vary the slope of the conveyor to enable goods to travel at a constant velocity. Roller conveyors are used, for example, to convey flat-based tote bins or pallets. The smoothness of the transfer depends on the size of the load in relation to the size of the roller. In some applications, the roller conveyor system is combined with bar code readers, which identify pallet, and programmable logic controllers or industrial computers to route pallets among a variety of destinations.

Belt

Belt conveyors (Figure 8.3(b)) use a flat belt made from a variety of materials to carry goods between destinations. Some belts are made from toughened plastic, while others are made from a metal mesh. The belt is looped around a driving roller at each end and is supported between each end by a frame structure and/or idling rollers. Materials are placed directly on the belt or are transferred in pallets and tote bins.

Chain

Chain conveyors (Figure 8.3(c)) use chains — very often similar to bicycle chains — to transport goods. The chain is driven by powered sprockets at various stages along the chain. One or more chains operate in parallel, transporting goods in tote bins or pallets. Sometimes fingers are attached to the chain to act as holders for transporting units.

Overhead

Overhead conveyors (Figure 8.3(d)) consist of a cable or chain routed along a path which links two or more destinations. Trolleys or hooks are attached to the cable at various intervals and goods are attached to these. The loads are suspended in the air as they travel between their destinations. The overhead cable uses a minimum of factory floor space and can be routed over long distances and looped around corners with relative ease.

Material handling conveyors are very common in industry and have a number of advantages. They are very reliable for transporting goods between destinations. They require low technology and are very dependable. In recent years, conveyor systems have become more sophisticated, as bar code systems are used to identify goods and route them between several destinations. Shuttle gates are frequently used to link conveyors while allowing passageways between conveyors. Figure 8.4 shows how a shuttle transfers goods across a short gap in the conveyor, provided nobody is passing through the gap.

FIGURE 8.4: SHUTTLE LINKING CONVEYORS

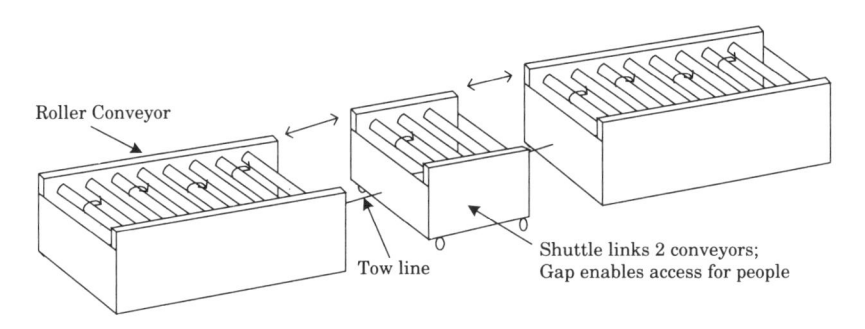

Roller Conveyor

Tow line

Shuttle links 2 conveyors; Gap enables access for people

There are also some disadvantages associated with using conveyor systems. They are a relatively rigid and inflexible means of transporting goods, and once the original purpose for which they were designed is obsolete, it can be difficult to reconfigure and adapt the conveyor for new tasks. Conveyors tend to be custom-designed for each application and the components of the system are not readily adaptable to new situations. It can be cumbersome and expensive to incorporate routing capability into conveyor systems, especially when a high number of destinations are involved. The destinations must be physically linked, which can lead to a maze of conveying equipment. Conveyors also tend to hold a lot of inventory. In the current climate of tight control on inventory levels, it can be difficult to control the level of inventory held within conveyor systems.

Automated Guided Vehicles

Automated guided vehicles (AGVs) overcome many of the problems associated with conveyors. AGVs are self-propelled vehicles, which often look like unmanned pallet trucks moving along aisleways, picking up pallets in one location and depositing them in another. They are powered by on-board rechargeable batteries and typically operate for 12 to 16 hours or more before needing to be recharged. AGVs are guided by sensors on the vehicles which follow either a wire embedded in the factory floor or patterns on the floor. Wire guidance involves embedding a wire approximately 1 cm into the factory floor. The AGV recognises the position of the wire and basically follows its path. The guidance system works by sending an electrical signal using a frequency generator through the wire; this signal is monitored by two on-board coil sensors. The sensors control the steering of the AGV by tracking the position of the wire.

FIGURE 8.5: AGV STEERING MECHANISM FOR WIRE GUIDANCE METHOD

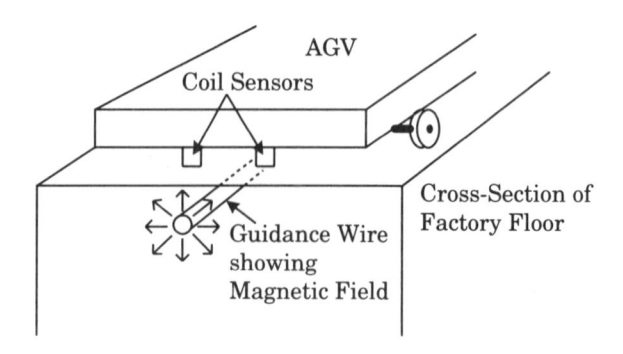

Another common tracking method is to use reflective paint on the factory floor. Light sensors on board the AGV control the steering by tracking the paint. It is important with this system that the paint is kept clean and unscratched. An alternative to paint is to tile the factory floor with black and white tiles similar to a chess board design. Light sensors detect variation in the tiles and calculate their route based on the size of the tiles.

Seeing AGVs for the first time can be intimidating. They appear to have a mind of their own as they move between locations. Safety features on AGVs are very important, to avoid crashing into other AGVs or injuring bystanders. Most AGVs are fitted with an emergency bumper that surrounds the chassis. If the bumper touches off any object during transit, the AGV brakes and powers down immediately. Other safety features include obstacle detection sensors such as infrared, optical or ultrasonic sensors at various positions around the AGV. When an obstacle is detected, the AGV either reduces speed or comes to a complete standstill until the obstacle is removed.

The management of AGVs is controlled in a variety of ways. On-board computers can direct the AGV to its next destination and make any necessary decisions at junction points en route. Routes can be predetermined or set by an operator after each load has been picked up or dropped. Some systems utilise remote call stations at load/unload points. These systems use infrared or radio signals to communicate with AGVs and request a vehicle to pick up a load. AGVs that are passing respond to the signal, pick up the load and receive instructions regarding delivery location. In large factories or warehouses, a central computer monitors the location of AGVs and plans their routes to maximise efficiency. At various stages along the routes, communication boxes provide real-time information regarding the location of AGVs, enabling the central computer to communicate pick-up requests.

AGVs have a number of advantages over conveyors. They are more flexible, goods can be carried loose, in tote pans or pallets, and they can be reprogrammed easily. In addition, they do not require a lot of floor space to carry goods, and it is easier to control inventory levels because of their lower capacity to store goods. However, in a simple environment where transport distances are short, they are more expensive than conveyors, and tend to need more engineering support during use for reprogramming and maintenance purposes.

ROBOTS

A robot is defined by the Robotics Industries Association as follows:

> An industrial robot is a reprogrammable, multifunctional manipulator designed to move materials, parts, tools, or special devices through variable programmed motions for the performance of a variety of tasks.

The definition basically covers a broad range of reprogrammable machines. The term robot was first coined in 1920–21 by the Czech playwright, Karel Capek, from the Czech word for forced labour or serf. Capek created the image of an automated programmable human. Modern industrial robots are not as sophisticated as Capek's image, but the term is mostly associated with machines that use an arm, wrist and gripper to grasp and manipulate materials. In general, robots are categorised into four common configurations, according to the structure of the arm. The purpose of the arm is to position the robot's wrist, which holds the gripper in a suitable position to enable the gripper to grasp and manipulate the object being processed.

FIGURE 8.6: ROBOT ARM CONFIGURATIONS

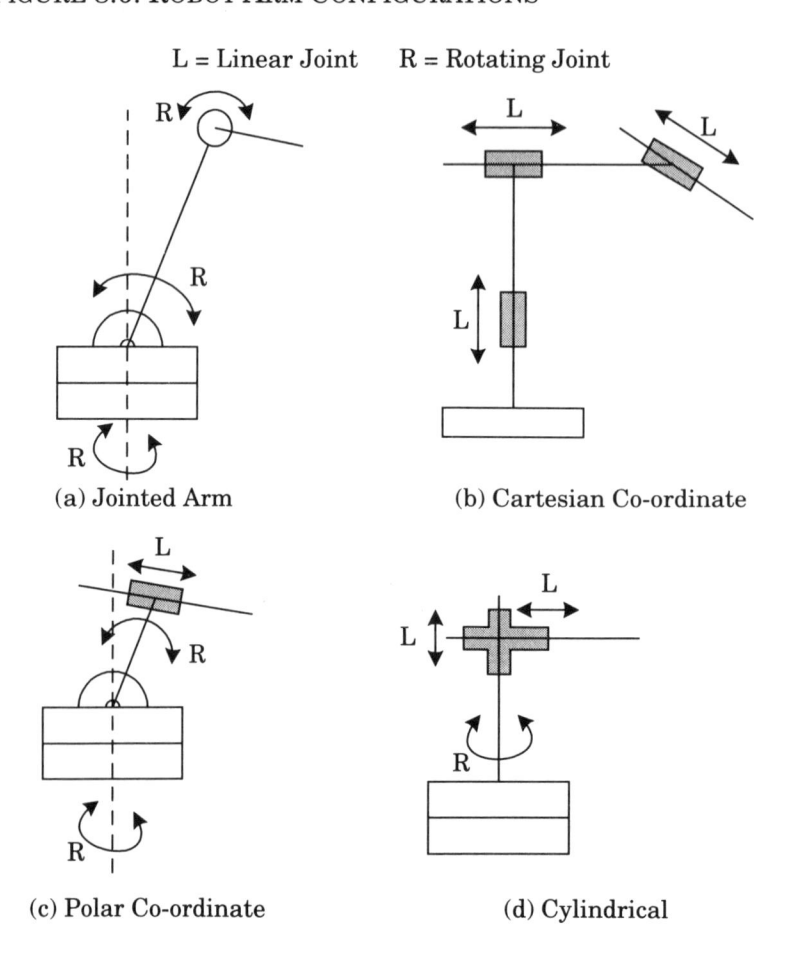

L = Linear Joint   R = Rotating Joint

(a) Jointed Arm

(b) Cartesian Co-ordinate

(c) Polar Co-ordinate

(d) Cylindrical

The jointed arm configuration (Figure 8.6(a)) has the general configuration of a human arm. The arm consists of a shoulder joint and an elbow joint, and can swivel around its base.

The Cartesian co-ordinate robot (Figure 8.6(b)) enables the arm to move in the x, y and z planes. It is commonly used to insert components in an assembly operation. It has three sliding joints which enable it to operate in a rectangular working space around the robot base.

The polar co-ordinate robot (Figure 8.6(c)) consists of a sliding joint and two rotational joints. The arm operates in a spherical space around the robot base.

The cylindrical robot (Figure 8.6(d)) consists of two sliding joints that swivel around the robot base. This configuration enables the robot to operate in a cylindrical space around the base. The horizontal columns move up and down the central column.

Attached to each arm is a wrist section, which may also have joints for further flexibility. The wrist may be attached directly through a rigid joint to the arm, or various sliding or rotating joints may be incorporated for increased manoeuvrability. The robot gripper or processing tool (end-effector) is attached directly to the wrist. The design of the wrist depends on the application and the end-effector used.

The type of end-effector used depends on the application the robot must perform. End-effectors are categorised as either tools or grippers. The robot can manipulates tools such as:

- Spray painting gun
- Welding tool
- Rotating spindles for drilling, routing, etc.
- Spot welding gun
- Assembly tools such as screwdrivers.

The robot control system directs the tool relative to the position of the work being processed. The robot must be able to transmit instructions to the tool that starts, stops and manages its actions.

Grippers use a variety of techniques — such as mechanical grippers and vacuum suction cups or nozzles — to grasp and manipulate material. Mechanical grippers pick up units by clasping them in a pincer movement similar to the way humans use the thumb and fingers. These grippers are custom-designed according to the mechanical features of the part being picked up. Vacuum grippers, by comparison, use suction cups or nozzles to pick up pieces. The size of the suction cup or nozzle orifice is dependent on the size and weight of the part being manipulated.

Rudimentary robots can be controlled by mechanical stops and/or limit switches for each joint. These methods limit the movement of the robot's joints between two defined positions and are used in simple pick-and-place systems. The robots have low accuracy levels and are often pneumatically actuated. Reprogramming consists of repositioning the mechanical stops to change the pick-and-place positions. Most industrial robots are more sophisticated than this rudimentary version and use a computer system combined with motors to control joint positions. These robots are programmed or taught tasks by means of a teach pendant or writing a program. A teach pendant is similar to the joystick used in computer games. It has toggle switches that control the movement of joints and the position of the end-effector. The operator uses the pendant to move the end-effector through a work cycle and the computer notes the position of joints at various positions during the work cycle in order to replay the actions later. There are two recording methods used: point-to-point and continuous path control. In point-to-point systems, the robot is taught specific locations in three-dimensional space in which to position its end-effector. The computer interpolates the distance between its current position and its next position in order to identify the path taken during movement. Robots with continuous path control operate in a similar manner to point-to-point control, except that during the teaching process, the computer records the *movement* of joints, which enables it to reproduce the transit path used during programming. Continuous path control is important when the end-effector is a tool such as a paint spray gun or router and the path used in moving between two points is critical. Continuous path control requires more computer power and memory than point-to-point control. Robots can also be programmed by using computer programs such as VALII, developed by Unimation; other robot manufacturers such as IBM, Toshiba and General Electric use their own languages but, in general, they have a similar approach and style. These languages use textual terms to instruct the robot:

- MOVE (variables) — indicate 3-D spatial position
- SPEED (variable) — indicates speed of movement
- WAIT (variables) — indicate condition that must be satisfied before stops waiting
- OPEN (variable) — indicates distance to open gripper
- CLOSE (variable) — indicates distance to close the gripper.

Teach pendants are easier for non-specialist personnel to use in reprogramming robots; however, direct programming with one of the above languages is more compatible with new CAM/CAD packages. Integration with these packages enables programs to be written without having to take the robot out of service. As software technology improves, these packages should become as easy as the teach pendant to use.

FLEXIBLE MANUFACTURING SYSTEMS

Flexible Manufacturing Systems (FMS) integrate many of the technologies already discussed into a single production system. As the name suggests, it involves a flexible production line capable of producing a wide range of products in small batch sizes. It consists of automated material handling equipment that delivers raw materials to and from work cells for processing. The work cells utilise robotic arms to load work pieces into machines for processing. The machines have automated tool change facilities that enable changeover between products to be almost instantaneous. Fast changeovers mean that small batch sizes as low as one unit can be processed. All the technologies within the FMS are controlled by a central computer system that monitors, schedules and controls all activities within the FMS. In effect, the ideal FMS is a completely automated production line.

FMSs are more common in the machine tool industry than in the assembly industries. Computer numerically controlled (CNC) machines have been developed for integration into FMS. CNC machine tools such as lathes, milling machines and drills have developed flexible systems that enable automatic tool changes. The various subsystems that constitute the FMS are controlled by the central computer. This computer supervises the workload of the various machines and monitors their performance; controls tool changeover and delivery of tools to the relevant workstations; and warns when a tool is reaching its end-of-life.

The central computer controls the movement of different batches within the system, routing them to machines as they become available. It also monitors the type of tools available at each workstation. If the required tool is not at a machine, the system notifies an operator to load the tool into the machine's tool magazine. During a tool change the magazine rotates into position for an automatic tool exchange. This quick changeover feature of CNC machines makes them ideally suited to FMS configurations.

Tool life is monitored by the controlling computer to minimise exposure to wear. If a tool has reached its end-of-life, the operator is notified by an alarm or flashing light and the tool is replaced in the machine's tool magazine. The central computer provides management reports that measure FMS utilisation rates, reliability of subsystems, production throughput, and tool usage rates, among others.

The machine tools within the FMS cannot hold an infinite variety of machine tools. Typically, each machine tool magazine holds between 5 and 20 tools. Group technology techniques (as described in Chapter 5) are used to create families of parts with similar machining characteristics. Parts are divided into primarily rotational or prismatic parts and subdivided according to other machining characteristics. FMSs tend to be designed for a particular family of parts with similar machining requirements. Using group technology minimises the extent of

Automation Technologies 273

changeovers between batches and the number of tools required within the system.

Figure 8.7 shows how an FMS configuration routes product between machines during manufacturing.

FIGURE 8.7: FMS CONFIGURATION WITH CONVEYORS

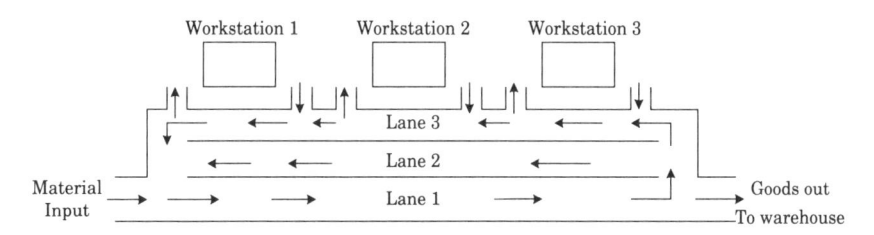

```
              Workstation 1      Workstation 2      Workstation 3
                 ┌────┐             ┌────┐             ┌────┐
                 │    │             │    │             │    │
              ↑↓ └────┘ ↑↓       ↑↓ └────┘ ↑↓       ↑↓ └────┘ ↑↓
           ┌──┘      └──┘ ←  ← Lane 3 ←  ←  ←  ←  ←  ←  ←  ←  ┐
           │         ←  ←       Lane 2       ←
  Material │                                                   → Goods out
  Input    →    →    →          Lane 1       →                 → To warehouse
```

This configuration uses a conveyor system to co-ordinate routing of parts between workstations. Parts enter the FMS conveyor from other areas within the factory and circle a buffer section of the conveyor until a workstation is free to process the parts. Parts are held on standard pallets which have bar code labels. Laser bar code readers at various stages along the conveyor identify the pallet to the FMS computer. A computer monitors and routes pallets of parts between stations. When a workstation is free, the parts are directed to that station where, in this scenario, an operator loads the part into the workstation. The processing program for the workstation is identified for the bar code on the part and/or pallet. After processing, the part is returned to its pallet and enters the conveyor system to be redirected to its next processing workstation. This cycle repeats until all operations are complete and the pallet exits the FMS.

Flexible manufacturing systems result in higher machine utilisation rates, reduced work-in-progress, higher flexibility in scheduling batches and a lower manufacturing lead-time. They are primarily used in the machine tool industry, but if products are designed to be compatible with automated assembly, flexible manufacturing systems will be developed for assembly lines with robots performing assembly tasks instead of operators.

AUTOMATED ASSEMBLY

Automated assembly is not as prevalent as automated parts production. There are several reasons for this, including the fact that some products may not have been originally designed for ease of automated assembly, or the high cost of automation may not be justifiable when compared to the number of units being assembled. As each of these issues is resolved, automated assembly becomes a viable option.

Traditionally, products have been designed for manual assembly. Operators are much more dextrous than robots and are more capable of carrying out complex tasks. Robots do not yet have the cognisance or the manipulative abilities of humans and therefore products have to be designed according to automated assembly guidelines. For example, an operator has little or no problem manipulating a screw, washer and nut to join materials, but these apparently simple tasks are difficult for robots. In this situation, a spot weld is perhaps more appropriate for automated assembly.

Designing for assembly guidelines are relatively well established; many companies have set up systems that ensure new products are suitable for automated assembly. When these guidelines are used, automation will become a viable option for medium to high volume production.

Design for Assembly Guidelines

Part Count Reduction

This involves questioning the purpose of each component of the assembly and investigating the feasibility of incorporating one component into another to reduce the parts count. For example, General Electric ask the questions:

- Is there relative movement between parts?
- Does the part require a different material?
- Does the part have to be removed for service?

When these questions can all be answered "no", then brainstorming sessions are used to incorporate the part into another component or eliminate it completely. A simple example is eliminating use of screws and nuts by using snap fits incorporated into the design of the housing.

Simpler Fastening Techniques

As the above example indicates, very often, screws and nuts can be replaced by snap fits. Other assembly friendly techniques include adhesive backed components, spot welding, crimping, soldering. These techniques reduce the need for manipulating several components simultaneously during the fastening process.

Design for a Straight Line Assembly

The assembly configuration should enable all components to be placed along the same axis. This is sometimes referred to as a "sandwich construction", where components are placed vertically in sequence.

FIGURE 8.8: DESIGN FOR STRAIGHT LINE MANUFACTURING

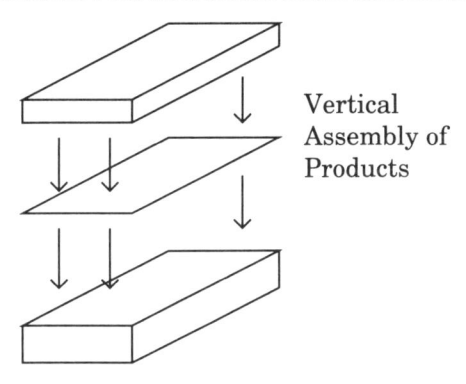

Vertical Assembly of Products

When it is not possible to design the product for vertical assembly, the number of directions from which components are joined should, where possible, be minimised. In this way, there is less flexibility demanded from the robot gripper.

Modular Design of Product

The product should be designed in modules, enabling different sub-assemblies to be constructed on separate machines. The modular approach reduces the risk of high equipment down time during production. If the unit is completely assembled at one workstation, a machine breakdown during any one task shuts down the complete system. The optimum number of assembly tasks to be performed at a workstation depends of the reliability of the equipment and the cycle time to produce the unit.

Positioning of Parts

Parts to be joined together should be designed for ease of positioning. They should be self-aligning with lead-in chamfers or guide pins. Parts should be designed symmetrically so that orientation is not important; when symmetrical design is not feasible, the shape of the component should indicate the placement orientation.

Component Quality

Components should be of a very high quality for automated assembly. Defective components cause high equipment downtime by jamming feeder tracks or causing incorrect positioning.

Presentation of Components and Sub-assemblies

Components and sub-assemblies have to be presented to the assembly robot in a definite location and, ideally, oriented correctly for direct placement. Three common methods are:

- *Magazines* in which components and sub-assemblies are stacked: the parts require load-bearing surfaces that eliminate the potential for damage, such as scratching.

- *Hopper compatible*: component supplied in bulk should be capable of being fed using a hopper and feed track. The hopper is a vibrating bowl with a spiralling ledge at the side along which the components are sorted as they travel. The hopper feeds a track that presents the component to the robot gripper.

- *Customised packaging*: used primarily in the electronics industry, where components are individually packaged in pocket tape and reel. As the reel is unwound, each pocket is opened enabling a vacuum nozzle to pick up the component.

The disadvantage here is that the cost of packaging often exceeds the cost of the component.

Component and Sub-assembly Design for Handling

When components and sub-assemblies are being designed, it is important to incorporate into the design a suitable place for the robot to pick up the unit. The pick-up zone should be compatible with the end-effector generally used by the robot to avoid having to change end-effectors for different components.

Design for Automated Inspection

The new product should be designed for automated inspection. Visual systems can be used to detect the presence of components or out-of-tolerance assembly work. The parts to be inspected should be within the field of vision of the camera and easily recognisable.

Design for Customisation

In the current market environment, the customer is demanding more options and choice in how their product is configured. Design for customisation involves designing the product to have a basic initial structure, to which extra options can be added at the later stages of assembly. In this way, customisation of the product occurs, in accordance with the customer's order, just before the product is shipped.

It is estimated that 50 per cent of assembly work is too low in volume to justify the expenditure on automated assembly. Other designs are too complex to be assembled with robots. Design for assembly techniques simplify the assembly requirements and places less demands on robots. The use of design for assembly guidelines and the development of less expensive robots will enable more manufacturers to consider automated assembly as a viable option.

COMPUTER INTEGRATED MANUFACTURING (CIM)

The term *computer integrated manufacturing* was coined by Dr Joseph Harrington to describe the integration of various manufacturing processes with computer technologies. Dr. Harrington pointed out that the integration of the computer technologies was more powerful than the sum of their individual parts and therefore greatly increased factory effectiveness. It is estimated that CIM factories can operate at break-even point at 30 per cent of operating capacity as opposed to 50 or 55 per cent for conventional factories.

CIM involves integrating technologies such as Flexible Manufacturing Systems (FMS), Computer Aided Design (CAD), Computer Aided Manufacturing (CAM), Computer Aided Process Planning (CAPP), Shop Floor Data Control (SFDC), Business databases for accounting and other computer systems, (an overview of these systems is provided in Chapter 7). By using a central controlling computer system linked to the various subsystems, data can be communicated to a central database and accessible to all other subsystems as required. In this way, the stock control system can monitor material levels in stores; the materials requirement planning (MRP) software analyses forecast demand and current stock levels and can place orders for raw materials as needed; and the accounts department monitors the receipt of goods in order to pay suppliers. Computer integrated manufacturing links these various sub-systems, enabling them to operate almost as a single system.

Computer integration strengthens the links between groups by enabling people to access central databases. Manufacturing engineers have access to new product designs, at an early stage in their development, and can ensure that these products are designed for ease of manufacture and assembly. The computer aided manufacturing packages can convert product designs into computer programs compatible with local machine control software. Machine programs can be stored in a central computer and downloaded to machines according to production schedule requirements. Computer integration enables different departments within the organisation to work more closely together and therefore increases the responsiveness of the company and its overall performance.

FIGURE 8.9: COMPUTER INTEGRATED TECHNOLOGY LINKS VARIOUS SOFTWARE SYSTEMS

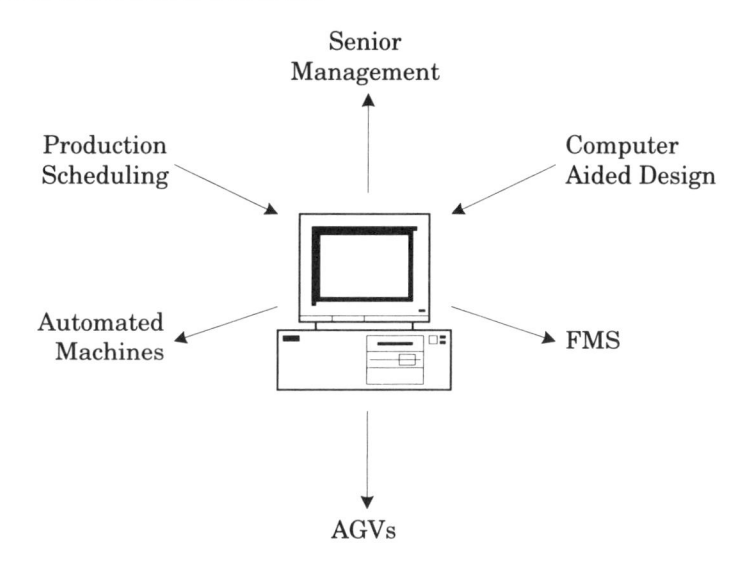

```
              Senior
            Management
                ▲
                │
Production ──▶ [PC] ◀── Computer
Scheduling       │      Aided Design
                 │
Automated ──▶  [PC]  ──▶ FMS
Machines         │
                 ▼
               AGVs
```

The technologies which enable CIM factories to become a reality already exist. Many engineers in factories are currently working to integrate the various islands of automation into the factory's central computer system. As software and communication protocols are developed, computer technologies will be easier to integrate, and CIM factories will become more prevalent. As companies evolve and adopt CIM technologies, they take on many of the characteristics associated with the image of the "factory of the future".

THE FACTORY OF THE FUTURE

A vision of the factory of the future helps to generate an integrated technology strategy. It summarises how all of the technologies described above combine to create what the Japanese call a "lights out" factory or an "unmanned" factory where all physical operator work is automated. The factory of the future will have to be able to compete on price, quality, speed of delivery, and new product options. It is a long time since people were happy to accept only black cars from Henry Ford. Therefore factories will have to be flexible and responsive to their customers. Factories of the future will probably have many of the following characteristics:

Fully Automated and Integrated Production Equipment

The high level of automation will require little if any labour to operate equipment. Production flow will be continuous by nature. There will be:

- Automatic routing of material between workstations
- Automatic equipment set up
- Information technology systems maintaining production databases and controlling the production system.

Small Lot Production and Short Lead Times

The trend towards customisation of product and reduction of inventory levels will lead to versatile machines which are capable of building small lots as required.

24-hour Operation

In order to pay for the high cost of equipment, 24-hour operation, seven days a week will be necessary.

Closer Control of Production by Marketing

Information technology is bringing marketing and production organisations closer together. These close links will enable the marketing department to better represent the immediate requirements of the customer and will also ensure that factory scheduling and organisation is focused on satisfying these requirements.

High technologies are expensive and not always as reliable as expected. Companies are cautious about technology investment and prefer to experiment with new technologies before making major investments. They are gradually changing their systems to align themselves with this image of manufacturing in the future, but it will be many decades yet before these factories become the norm.

In this image of the factory of the future, human workers will still be present. They will be highly skilled in equipment maintenance since any breakdown could potentially shut down the entire factory. Continuous improvement projects will attempt to eliminate such problems. Engineers and software programmers will be introducing new products, updating software systems and improving existing processes on an ongoing basis. There will also be a requirement for the personal touch in dealing with the external world. People will continue to work in receiving and shipping departments and in other less production-focused environments; thus it will be many decades, if indeed ever, that this level of the "unmanned" factory becomes a reality.

SUITABILITY OF AUTOMATION TECHNOLOGIES

Medium-volume batch manufacturing industries have the most to gain from developments in automation technologies. High-volume mass manufacturers use dedicated transfer lines, which were developed by Henry Ford, to minimise manufacturing costs per unit. Low-volume manufacturers will probably continue to use isolated machines working independently to produce goods. They cannot afford the high investment required for automation. However, medium-volume batch manufacturers with a diverse product range need flexible systems to cope with manufacturing complexities. Integrated computer systems reduce the time to market for new products. Flexible manufacturing systems control production scheduling, minimise machine set-up times, and improve equipment utilisation levels. Figure 8.10 shows the relationship of different technologies to different categories of manufacturing.

Despite the suitability of automation technologies to batch manufacturing, there has been no universal adoption of the new technologies. There was a high rate of investment in automation technologies in the early 1980s but demand for these technologies declined as the decade came to a close. It is

FIGURE 8.10: SUITABILITY OF AUTOMATION

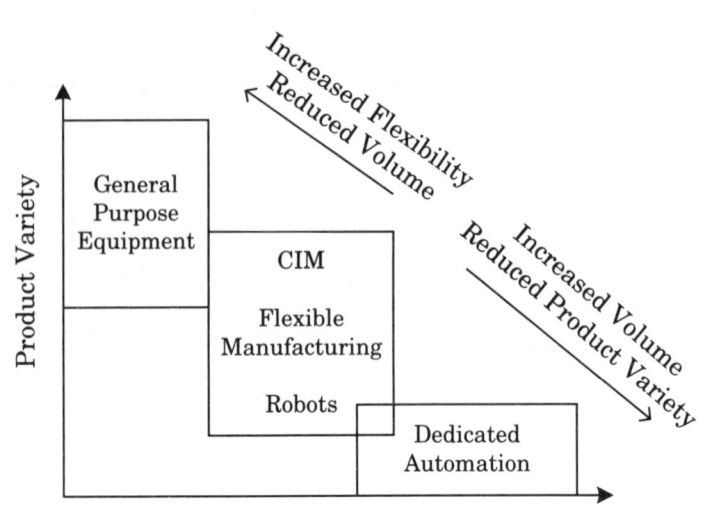

```
         ▲
         │         Increased Flexibility
         │        ← Reduced Volume
         │  ┌─────────────┐
         │  │  General    │
Product  │  │  Purpose    │
Variety  │  │  Equipment  ┌──────────────┐
         │  │             │  CIM         │         Increased Volume
         │  │             │              │        Reduced Product Variety
         │  │             │  Flexible    │
         │  │             │  Manufacturing│
         │  │             │              │
         │  └─────────────┤  Robots  ┌──────────────┐
         │                │          │  Dedicated   │
         │                └──────────┤  Automation  │
         │                           │              │
         │                           └──────────────┘
         └─────────────────────────────────────────────▶
              Volume in units of parts or products
```

only in recent years that new technologies are again experiencing high levels of investment. The reason for the variation in investment patterns is best summarised by Tom Forester (1989):

> People got carried away with the utopian visions of automated factories, overlooking the high cost of high tech and the enormous complexity of factory operations. Robots were absurdly over-hyped: it was conveniently ignored that they were both much more expensive and less flexible than humans. As one commentator puts it, "Contrary to the early hype, it rarely makes business sense simply to replace a human worker with a robot and expect the machine to pay for itself in saved labour costs". Much can be achieved by improving quality, and product or inventory flow without resorting to this expensive high tech "fix" . . . a total machine take-over in factories no longer seems to be the goal. Rather, it is a common-sense *partnership* between machine and man.

Tom Forester's explanation describes why companies are more cautious nowadays with regard to investing in technology. A report by the Ingersoll Engineers (Mortimer, 1985) who are leading consultants in manufacturing technology stated:

> This report advocates that integrated manufacturing is approached, not as a combination of existing or known technologies and equipment that must be linked together, but rather from first principles, by examining all aspects of a total business — its future, markets, people, manufacturing and products — and combining them in a new way to meet long-term business objectives.

Many companies attempted to introduce the new technologies into their current work environment, only to find that the systems present were incompatible. Manufacturing managers re-examined how their systems were organised and concentrated on simplifying them. In the latter half of the 1980s and the beginning of the 1990s, many manufacturing managers concentrated on projects like Just-in-Time, Total Quality Management and Total Productive Maintenance. These projects streamlined their production system and enabled many to turn again to technology as a means of improving performance.

Automation has the capability to achieve the goals for which it was designed, but only if it is supported by the work environment. Operators must be properly trained to support the new technologies; engineers must be able to integrate the new systems with the old. Today's manufacturing managers are taking a more prudent and long-term approach to technology investments.

TECHNOLOGY STRATEGY

Global competition means that companies are competing on a range of criteria including responsiveness, product variety, cost and quality. Automation technologies can improve performance in each of these areas, but success depends on a strategic, focused approach to technology investment. Companies wishing to improve responsiveness may emphasise investment in technologies that improve flexibility; other companies may be more concerned with productivity and concentrate on developing special purpose high-volume equipment. Companies can succeed in gaining competitive advantage through automation by having a focused strategy that ensures their leadership in a chosen area, such as flexibility, quality, responsiveness, or productivity. There are several approaches to technology investment, the suitability of which depends on the business circumstances prevalent within the company and the market. Groover (1987) identifies several different technology strategies, some of which are described below.

Faster, High-Tech, Special Purpose Machines

This involves developing customised machines to perform specialised tasks with higher throughput rates. Many western mass producers are attempting to compete by developing customised high-volume equipment that can match the productivity levels achieved by low-cost labour in developing countries.

Combining Tasks so that they are Performed by the Same Machine

Some companies concentrate on developing machines that can perform a sequence of tasks. This reduces the time lost in transferring parts between machines and therefore the overall manufacturing time. Machines now exist that enable turning, milling and drilling to be performed on the one set-up. As machine and processing technologies become more reliable, some equipment manufacturers are competing by producing machines capable of performing several tasks in the production process.

Simultaneous Operations

This involves performing two or more tasks simultaneously on the same machine. It is sometimes possible to perform tasks such as part labelling at the same time as the part is being tested. The result is a reduction in material handling and processing time.

Increasing Flexibility

Companies are achieving increased flexibility by automating product changeovers. Zero changeover time means that batch sizes as small as a single unit become viable. The company can reduce its minimum order quantity or increase its product range with no loss in performance efficiency.

Improved Material Handling and Storage

Automated material handling routes raw materials through the various workstations and tends to reduce work-in-progress, leading to shorter manufacturing lead times.

Computer Integrated Manufacturing (CIM)

As discussed earlier, computer integrated manufacturing aims to integrate the various factory operations. A central computer database links the various applications used to schedule production, design new products, and manage accounts, among others. CIM technology reduces the overall manufacturing lead time, reduces the time taken to develop new products and improves equipment utilisation.

Competitive advantage in manufacturing is achieved by having what your competitors wish they had. If a company limits itself to purchasing standard equipment, any advantage gained will easily be copied by its competitors. An alternative approach is to enter partnership arrangements with equipment manufacturers to develop proprietary manufacturing equipment, or else to nurture, in-house, the engineering talent that can custom-build machines. The first option involves surveying machinery manufacturers and identifying those who are capable and willing to design and manufacture machines which will advance one of the strategies already described, such as increased flexibility or combined operations. Ideally, agreements between the two companies prevent sale of the technology to competitors. The second approach — building machines internally — is frequently adopted by major multinationals. Many Japanese and European manufacturers develop their own machines. Companies like Siemens and Philips in Europe and Matsushita

Electric and Hitachi in Japan develop their own production equipment. Their leadership in machine design enables them to continuously improve their productivity and quality levels.

IMPLEMENTING AUTOMATION TECHNOLOGIES

Carol Beatty, in a study (1992) into the implementation of advanced manufacturing technologies, identified three key criteria for the successful implementation of new manufacturing technologies:

- Develop an effective project champion
- Plan for system integration
- Use organisational integrational techniques.

Beatty studied the attempts that companies made to implement modern technologies over a three-year period and demonstrated how the lack of any one of these factors greatly reduces the prospect of success.

Each of the successful companies in her study had an effective champion who was familiar with the capabilities of the new technology and how it could improve current manufacturing. These people are the driving force behind the project. They sell their ideas to top management and obtain the resources and support needed to introduce the technology.

Introducing manufacturing technologies is not a solo effort on behalf of the project champion. Operators, technicians and engineers will have to use the new technology as a tool to make the process and the operation more effective. The champion must be a team player able to motivate work colleagues, listen to problems and take realistic steps to overcome obstacles. The champion needs people skills that enable them to delegate responsibilities and manage the implementation process.

It can be difficult to find suitable champions with all the skills required to manage technology projects. Engineers understand the capability of new technologies, but often lack people management skills. Companies need to invest more time and effort in providing training for these people, not just in technology but also in people management skills. Turning a vision of automation into reality requires competency in both areas.

Many technology manufacturers develop their software to support their products' individual needs. This frequently leads to various systems from different technology vendors being incompatible, and the opportunity for integration between systems is lost. Planning for system integration at the outset of the project minimises exposure to investment in systems that are incompatible. The problem is the lack of clear standards or universally accepted protocols for communication between software packages. Many of the potential benefits derived from modern technologies are lost when systems cannot communicate interactively. Integrating systems means creating an environment similar to CIM where people from different departments have access to the same data. The team must consider how the various systems will communicate, and perhaps only purchase from vendors who can satisfy generic requirements.

The third factor in Beatty's analysis is organisational integration. If people from various departments need to use a common database, the characteristics of the system will have to satisfy a range of cross-functional needs. The implementation project will need cross-functional teams and structures to design and implement the new system. For example, if the materials department are setting up a new inventory management software on the production floor, it will have to be user-friendly for operators to use; it will have to be compatible with existing systems within the MIS department; it will have to provide accurate reports for the production schedulers, the purchasing and finance departments. Each department will have expectations of the new system; success depends on teamwork and co-operation across departmental boundaries.

New technologies tend to introduce elements of uncertainty into the production environment. High levels of equipment downtime often occur during the introduction stage of a project such as the implementation of an FMS or an automated material handling system. Many companies have never overcome these early problems and have scrapped millions of pounds of equipment. Other companies have had to commit excessive engineering resources to sustain the technology without ever fully recouping their investment. Companies attempting to introduce new manufacturing technologies often need to reexamine how they manage technology on the production floor. Some experts accredit Japanese success at introducing new technologies to their investment in training their operators and technicians. Machine operators are able to supervise the new technologies and make minor repairs while technicians sustain and improve the performance of their machines. Both operators and technicians partake in quality circles to develop new solutions to equipment problems. This educated focus on maintaining equipment on the factory floor is a major element in making technology work. In western manufacturing, self-directed work teams (levels 3 and 4 operator empowerment — see Chapter 1) provide a suitable structure for ensuring a fast response to equipment problems. Members of the self-directed work team are empowered to make decisions and resolve problems as they arise. They provide engineering staff with much information regarding the possible causes of equipment downtime and their support becomes the vital link in ensuring the successful introduction of new automated equipment.

One company in which I worked wanted to enter a fast-growing market with a product designed for high productivity, capable of competing with production centres in China and the Middle East. The new product was a major advance on the company's current product range. The aim of the project was to be in high volume manufacturing in 1.5–2 years with a new product. The company succeeded in:

- Designing the new product for ease of assembly;
- Setting up a pilot production area that supplied customer orders. The capacity of the pilot line was to be approximately 10 per cent of the anticipated full production volume;
- Designed a special purpose high-volume production machine that combined many of the tasks in the prototype line and reached its high-volume target.

The major drawback was that the original schedule slipped by two years. It took approximately four years from product conception to resolving the teething problems with the high-volume equipment. Overall, the project has to be considered a success in that it introduced a new product with very high potential sales to the company. If we look at the framework behind this success, many of Beatty's ideas are evident.

The project was initially championed by an engineer with over ten years' experience in product design and production processes. He reported directly to the vice-president in charge of the project and was responsible for product design, setting up the pilot production line and developing the high-volume production equipment. The engineer was able to use a variety of consultants to advise on new product and equipment design. During these initial stages, most of the work was, by far, technically orientated, involving meetings with outside consultants. There was minimal need for the champion either to sell the project within the organisation or to manage a group of internal engineering staff.

The promoters of the project were the marketing department and the vice-president of the production facility. They ensured that the finance was available for the project and provided advice and assistance as needed.

When the project was developed to pilot stage, a second champion was hired to convert the pilot line to the high-volume line. This person had the production management and technical skills required to hire the people, train them, arrange the production systems, and interface with engineers on technical developments.

The technical champion started out as leader of the project, putting in the time and effort to kick-start the project. The concept of the ideal champion is as persuader and people manager. These roles were filled by two fellow-employees: the vice president and the new production manager.

The second factor influencing successful implementation was the integration of systems. In this example, integration of equipment hardware was more important than integration of software systems. Integration was overcome by purchasing much of the tooling equipment from the same supplier. In this way, the equipment manufacturer was unable to blame his competitors' equipment in upstream processes for the poor performance of his own. This policy led to a big reduction in problems associated with tooling equipment.

The third component of successful implementation is organisational integration. The level of integration required between the new production line and other lines was minimal, as resources were available to hire dedicated operators, technicians and engineers for the new line. It used the existing data management system and the same support departments, such as human resources and facilities. Other than that, it was relatively self-sustaining. Integration within the factory was aided by the vice-president who had responsibility for the factory and for promoting the project.

Successful implementation of new manufacturing technology can be difficult. The success of the project depends on the skills of the project champion, the level of integration between the various elements of the new system and the ability to integrate the new technology into the current organisation. When the technology is new and untried, perseverance may be another factor separating success and failure. While it may not be easy, successful implementation can be the difference between following the competition and leading the field.

CONCLUSION

Automation technologies can potentially enable western manufacturers to regain much of the competitive advantage they have lost. Companies from Asia already have the advantage of a plentiful supply of low-cost labour, which enables them to achieve high productivity levels. It is only through utilising automation technologies that companies can fend off these rivals and regain marketshare.

Investment in automation will only accelerate when manufacturing management appreciate the capabilities of these new technologies. This chapter provides an overview of technologies such as machine vision systems, robots and flexible manufacturing systems, among others. From this overview, it is possible to predict the nature of the factory of the future. Success in the next century will depend on management's ability to make wise investments that progressively move the company towards this futuristic vision.

Making a wise investment decision is only the first stage of technology management. Beatty's study shows that successful implementation needs a strong project champion with the people skills and technical skills to manage the project to completion. In the exhilarating world of high technology, it is easy to discard the need for integration of the new system with existing systems, and integration of the new technology into the overall organisation. These are two factors critical to achieving an adequate return in investment. Integration of different aspects of the factory is a key advantage of technology, because it enables the organisation to act as a unified entity.

Those who invest in these technologies will lead in areas such as flexibility, product cost, quality and responsiveness. They will dominate the market and ensure the demise of smaller competitors. Technology strategy has to be planned today in order to ensure a future for the company in tomorrow's marketplace.

Chapter 9

PROJECT MANAGEMENT

INTRODUCTION

The manufacturing world is changing at a faster rate than ever before. Management and engineers are coping with new technologies, shorter product life cycles and international competitors. In this world of change, manufacturing managers must be able to react quickly to new developments. They are continuously upgrading production facilities to take advantage of modern developments as they arise.

Project management involves applying a systematic approach to achieving the goals of change. When the concepts associated with project management are applied, the probability of successful implementation is greatly improved. These concepts offer the practitioner a step-by-step guide to project planning.

The techniques described in this chapter include networking tools such as critical path analysis, which identifies the duration of the project and its critical tasks, and the Gantt chart, which is used to represent project tasks and their associated duration time. The Gantt chart and resource histograms are used to optimise the utilisation of resources. Project management also involves managing the people and circumstances affected by change. Lewin's ideas on force-field analysis and the three stages of change (unfreeze, change, refreeze) are introduced.

A systematic approach to planning projects minimises the risk of oversight, improves project credibility and aids in the fast implementation of change. The techniques of project management are easy to learn. They are powerful tools for managers, and engineers who have responsibility for introducing change. They enable the project manager to predict resource requirements and the overall project duration.

DEFINITION OF PROJECT MANAGEMENT

ISO 8402 defines a project as:

> a unique process, consisting of a set of co-ordinated and controlled activities with start and finish dates, undertaken to achieve an objective conforming to specific requirements including constraints of time, cost and resources.

Project management refers to the administration of the project, its supervision and organisation. It means analysing the objectives of the project, defining the tasks needed to achieve these objectives and controlling their execution.

SUITABILITY

The techniques that will be presented in this chapter are applicable to a broad range of projects, and the merit of each technique will depend on the complexities involved in the project. Large-scale projects will tend to use each technique described while smaller ones may utilise only one.

For example, introducing hardware into the information services department may affect only a small group of people and may be easily implemented with a minimum of project management. However, if the shop floor data collection system is being changed, it may be necessary to gain inter-departmental support for the changes; to programme the alterations; to arrange training courses for operators who will be responsible for introducing the new system; and to verify that the new system is used in practice. Critical path analysis and resource optimisation are used to ensure that the changes are introduced on time. Lewin's force-field analysis and three stages to change are used to minimise resistance to the project.

Developments in computer software make it easier to use project management techniques and therefore possible to perform "what if" analysis on projects. These software tools mean that development of project networks and Gantt charts is relatively easy. The project team can experiment with changing the sequence of events or changing the duration times and observe the effects on project costs. Software packages make project management techniques amenable to a wider group of people.

DEFINING THE PROJECT

The first step in any project is to have a clear understanding of the goals of the project. This means understanding the needs and expectations of the project's sponsor or customer. It is important to verify the need for the project, that the project description is the best way of satisfying the customer's needs, that resources will be available to satisfy that need, and so on. This initial phase of the project's life ensures the relevance of the project and that support is available for implementation.

For example, the goal may be to introduce a new shop floor inventory management system in order to improve inventory utilisation. However, a JIT program or investment in the bottleneck process may also achieve these underlying objectives. The project manager verifies that the project's objectives are suitable to the customer's overall ambitions.

This phase of the project should result in clear, measurable objectives. Resources are allocated to achieving these objectives within an agreeable time-frame. The next step in the project is to communicate the objectives to those who will have to develop and implement the project tasks, and to those who will be affected by implementation. Those involved in the implementation process must understand how they can contribute.

For example, a general manager of a multinational subsidiary wanted to be the first within the corporation to obtain ISO 9000 accreditation. The award would provide the subsidiary with much prestige within the organisation and at the same time improve controls within the factory. A meeting was organised with all employees to communicate the goal. An outline was given of what was involved in achieving ISO 9000 and what the resulting benefits would be for the subsidiary. A temporary employee had been hired on contract for the sole purpose of championing the project and over the next few days, he provided more information and clarification for each employee.

This example shows how a clear objective is established and resources allocated to the project. All people affected by the project have been informed of its content and should be able to contribute to its successful completion.

Choosing appropriate goals is a critical initial step in implementing change. The project's chance of success is also much greater when the need for attaining the goal is self-evident and has the support of all concerned. The second step is to communicate the importance of the project to those who will have to convert the aspirations of the project into reality.

ANALYSIS OF PROJECT MANAGEMENT TECHNIQUES

Identifying Project Tasks

The team must identify the tasks that will enable them to achieve their goal. When similar projects have been completed previously, the team may be able to adopt many of the tasks carried out by this earlier team, but when the project is new and novel, there is usually uncertainty regarding the work required and the solutions needed.

Brainstorming, as described in Chapter 2, is a technique for generating creative ways of achieving project goals. During the project meeting, the team generates ideas and possible alternative solutions to achieving goals and objectives. At the end of the session, the team agrees on the best means to achieve the goals and lists the tasks associated with this approach.

The work breakdown structure (WBS) is another method of defining how to achieve the project's objectives. WBS involves identifying the major high-level tasks to be achieved in order for the overall project goals to be met. For example, the team may identify their primary objective as "to improve the performance of a machine by 15 per cent". The team defines the following criteria as high-level requirements for achieving their goal:

- Fully trained operators

- An effective preventative maintenance system

- Tight control on raw material quality.

The team now breaks down each of these items into their basic component parts so that all the tasks needed to achieve each of the criteria are identified. As the tasks are completed, the criteria or high-level tasks are achieved and the overall project goals are attained.

FIGURE 9.1: WORK BREAKDOWN STRUCTURE

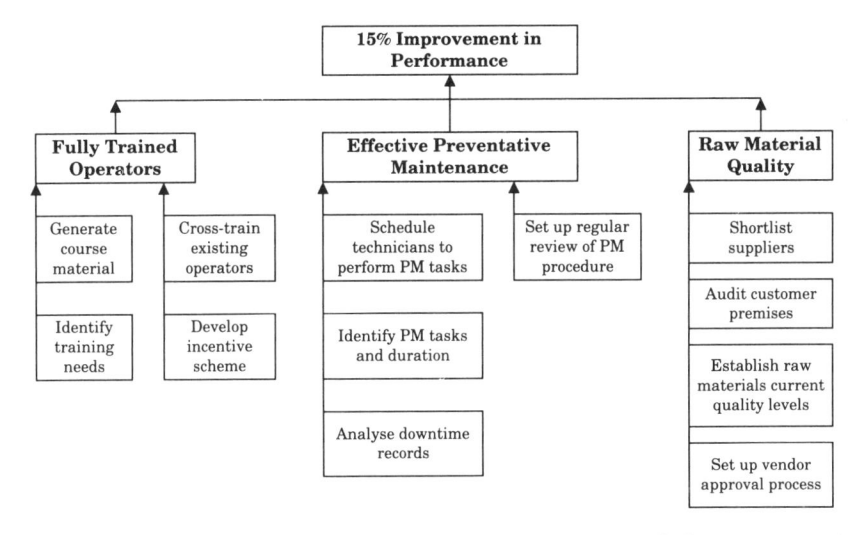

```
                        ┌─────────────────────┐
                        │ 15% Improvement in  │
                        │    Performance      │
                        └─────────────────────┘
                                  ▲
          ┌───────────────────────┼───────────────────────┐
┌──────────────────┐    ┌──────────────────────┐    ┌──────────────┐
│  Fully Trained   │    │ Effective Preventative│   │ Raw Material │
│    Operators     │    │     Maintenance       │   │   Quality    │
└──────────────────┘    └──────────────────────┘    └──────────────┘
```

- Generate course material
- Cross-train existing operators
- Schedule technicians to perform PM tasks
- Set up regular review of PM procedure
- Shortlist suppliers

- Identify training needs
- Develop incentive scheme
- Identify PM tasks and duration
- Audit customer premises
- Establish raw materials current quality levels

- Analyse downtime records
- Set up vendor approval process

Sometimes the WBS helps to distribute responsibilities to various team members. The WBS often breaks the project into groups so that each member of the team can take responsibility for a high-level task. For example, in the description of the machine improvement project, the maintenance manager takes responsibility for ensuring an effective preventative maintenance system, the production manager takes responsibility for ensuring operators are trained and the quality manager ensures that high quality raw materials are received.

TABLE 8.1: TASKS AND RESPONSIBILITIES

High Level Tasks	Responsibility
Fully Trained Operators	Production Manager
Effective Preventative Maintenance	Engineering Manager
Tight control on raw material quality	Quality Manager

Networking to Show Relationships between Tasks

Once the team has identified the tasks needed, they must then identify the relationships between the tasks, the sequence in which each task is performed and the interdependencies between tasks. Some tasks will have to be completed before others can begin while some other tasks can be completed in parallel. These relationships are identified by drawing a network diagram.

A network diagram is relatively easy to draw. Each task is listed on an individual slip of paper, as in the example below. The slips of paper are pinned on a white board to represent the sequence in which tasks take place. Lines and arrows are drawn between the slips of paper to show which tasks follow on from others.

FIGURE 9.2: NETWORK DIAGRAM

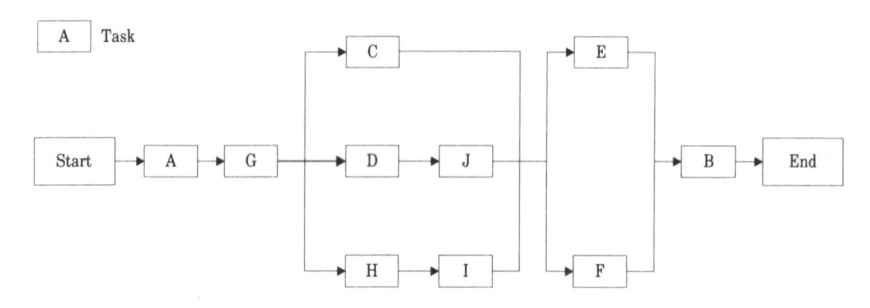

Drawing a network diagram is an iterative process as the team explores the relationships that define how each task relates to the others. The network diagram aims to portray how the various tasks relate to one another — which tasks have to be completed before others begin, which can be performed simultaneously, etc.

Estimating Task Duration Times

There may be a temptation to use rough subjective opinions regarding the time needed to complete the various tasks. This approach will greatly negate any benefits from project planning. The team should take time to research estimated duration times per task.

The task durations can be estimated using historical data from similar projects, or consultants/experts with prior experience may be brought in to advise the team.

The team can also use the PERT (Programme Evaluation and Review Technique) system for estimating the average or mean duration time for tasks. The team provides a best guess estimate from any information available. They then identify potential risks or good fortune that may affect the project. From this a pessimistic and optimistic estimate of duration times are generated. The PERT system calculates the estimated mean duration time as:

$$\text{Average or mean duration time} = \frac{(P + 4 \times BG + O)}{6}$$

Where P = Pessimistic, BG = Best Guess, O = Optimistic

The PERT estimation system enables the team to reflect some of the variation in task duration that occurs in real life. For example, in travelling to work each day, travel time varies depending on the traffic, on the level of road works being undertaken, on the initial starting time, and other such variables. The PERT calculation attempts to factor in these variations.

Critical Path Analysis and Project Duration

The task duration can be reflected on the network diagram by adding boxes to each task. The following figure shows a network diagram with the duration times added. Extra boxes have been added which will be fulfilled by critical path analysis techniques. Again, each letter reflects a task.

FIGURE 9.3: NETWORK DIAGRAM WITH TASK DURATION TIMES

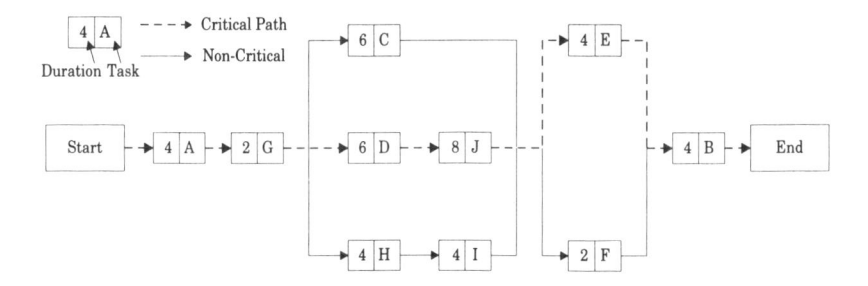

The critical path is the sequence of tasks that have the longest total completion time. It highlights the series of tasks that control the earliest finish time for the project. The critical path in the above example is AGDJEB, with a total duration time of 28 weeks. The shorter time required for these paths means that each path has spare time as they await completion of tasks on the critical path. Other paths with shorter total durations are:

	Duration	Spare time
AGCEB	20	8
AGCFB	18	10
AGHIEB	22	6
AGHIFB	20	8

This spare time can be linked as floating time to the various tasks to indicate that certain tasks can be delayed in starting with no negative affect on the overall project duration time. This can be very useful information because, when we look at resource utilisation, the float time available to tasks can be used to schedule non-critical tasks in order to reduce peak demand on key resources. For example, if there is a peak demand for 8 electricians during week 5 of a project and only 6 electricians are available, then tasks with floating time which require electricians during this period can, potentially, be rescheduled to reduce peak demand to 6 electricians.

Calculating the Float Time

The first step is to document on the network diagram the early start and early finish times for each task. An early start time equal to zero is entered for the first task. The early finish is equal to the early start plus the task duration. The early start of the subsequent task is equal to the early finish of its predecessor. Figure 9.4 shows the example.

The times shown indicate the earliest that each of the tasks can be started and finished. The earliest each task can begin is determined by the earliest finish time of its predecessor. Tasks E and F above can only begin once tasks C, J and I are

Project Management   301

FIGURE 9.4: EARLY START AND EARLY FINISH TIMES

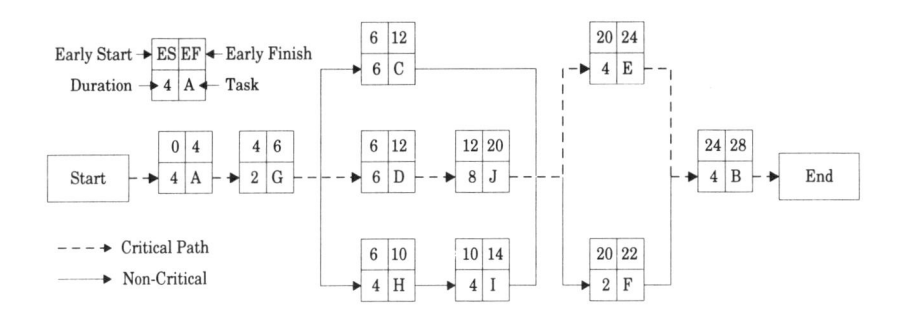

```
Early Start → ES EF ← Early Finish           6 | 12                    20 | 24
   Duration → 4  A ← Task                     6 | C                      4 | E

                    0|4    4|6    6|12    12|20            24|28
       Start    →   4|A →  2|G -→ 6|D  →  8|J   --------→  4|B  →  End

    ---→ Critical Path                 6|10   10|14    20|22
    ———→ Non-Critical                  4|H  → 4|I   →  2|F
```

completed. Therefore, the earliest start for tasks E and F is the earliest finish for tasks C, J and I. Since task J has the longest duration, its early finish time provides the early start time for task E and F.

The late start and late finish for each task which do not affect the project duration, are determined in a similar fashion, but this time, calculations are made by moving backwards from the project end task. The late finish for task B is equal to the finish for the project. Therefore the late start is equal to the late finish minus the task duration time. For B, the late start is 24. Note that because B is on the critical path, the early start is equal to the late start and early finish is equal to the late finish. There is no scope for any slip in time without affecting the overall project duration. The late finish and late start can be successively calculated for each of the other tasks

FIGURE 9.5: LATE START AND LATE FINISH TIMES

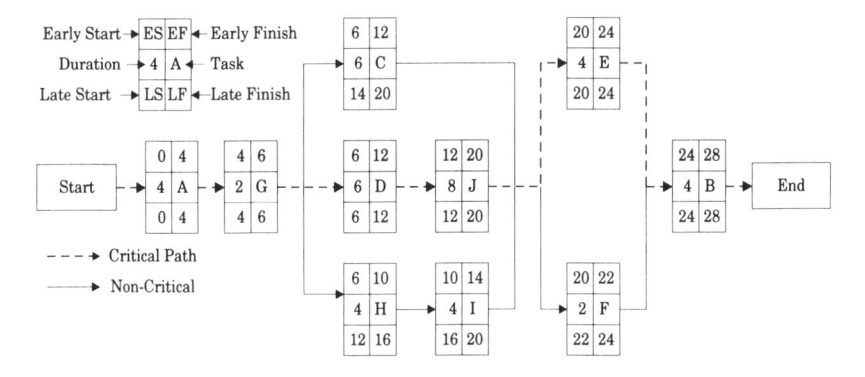

```
Early Start → ES EF ← Early Finish    6 | 12                  20 | 24
    Duration → 4  A ← Task             6 | C                    4 | E
   Late Start → LS LF ← Late Finish   14 | 20                  20 | 24

                   0|4   4|6    6|12   12|20          24|28
      Start    →   4|A → 2|G -→ 6|D →  8|J  -------→  4|B  → End
                   0|4   4|6    6|12   12|20          24|28

   ---→ Critical Path             6|10    10|14    20|22
   ———→ Non-Critical               4|H  →  4|I   →  2|F
                                  12|16   16|20    22|24
```

by progressively working backwards through the network diagram. For example the late finish for F is equal to the late start for G. The late start for F is equal to the late finish minus the task duration time, i.e. LS(F) = 22.

The floating time available to each non-critical task is the free time available for varying the start or finish of the task without affecting the overall duration of the project. The floating time is calculated as the difference between the late finish and early finish times. Figure 9.6 shows the float times calculated for each task.

FIGURE 9.6: FLOAT TIMES FOR THE VARIOUS TASKS

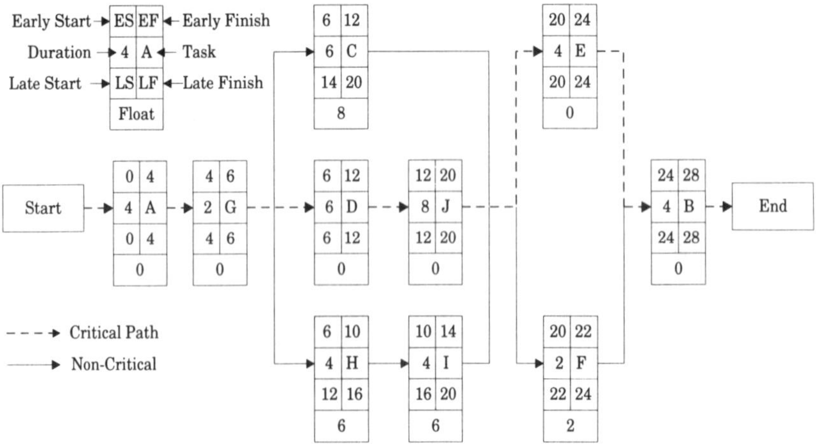

A floating time exists only for tasks not on the critical path of the project. It means, for example, the start time for task F can vary by two weeks without affecting the project completion date. Tasks C and K have the same float time associated with them — eight weeks. The float is linked to both tasks and must be shared. The float time for tasks H and I must also be shared. Floating time defines the spare time available for starting tasks.

Project Management   303

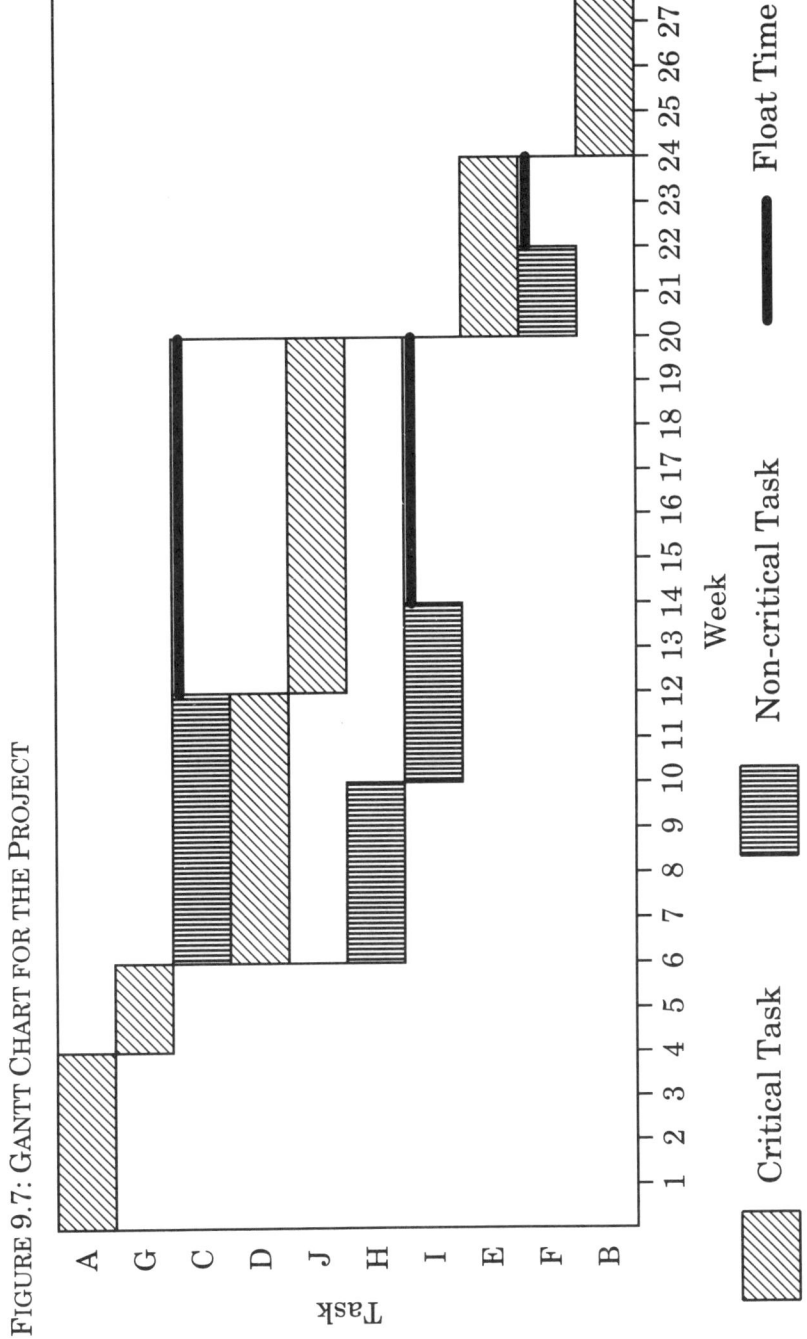

FIGURE 9.7: GANTT CHART FOR THE PROJECT

A Gantt Chart

A Gantt chart provides a convenient and powerful means of representing much of the information derived from the critical path analysis. Gantt charts are a graphical representation describing the sequence in which tasks are to be performed. All tasks are listed on the Y-axis and the X-axis is the timeline. Figure 9.7 shows the network from the example converted into a Gantt Chart.

Linkages between tasks can be portrayed by thin or coloured lines linking the bars. The floating time can also be represented; in this example, a thick line is used to represent the float time available. The critical path in this example is represented by diagonally shaded bars and non-critical tasks are represented by vertically shaded bars.

The Gantt chart presents the information is an easily digestible format. It also enables the team to begin evaluating the resources needed for the project.

Resource Utilisation

There may be a variety of resources required to complete the project — for example, time, cost, labour. Inevitably, resources tend to be limited in availability and the team needs to identify the various levels of utilisation throughout the project. If the critical resource is labour, then the utilisation of labour can be plotted for the period of the project. For the tasks in the earlier example, the following labour resources are needed:

Task	Technician Requirement (Labour)
A	0
G	2
C	3
D	1
J	2
H	4
I	2
E	2
F	1
B	0

FIGURE 9.8: GANTT CHART AND RESOURCE UTILISATION

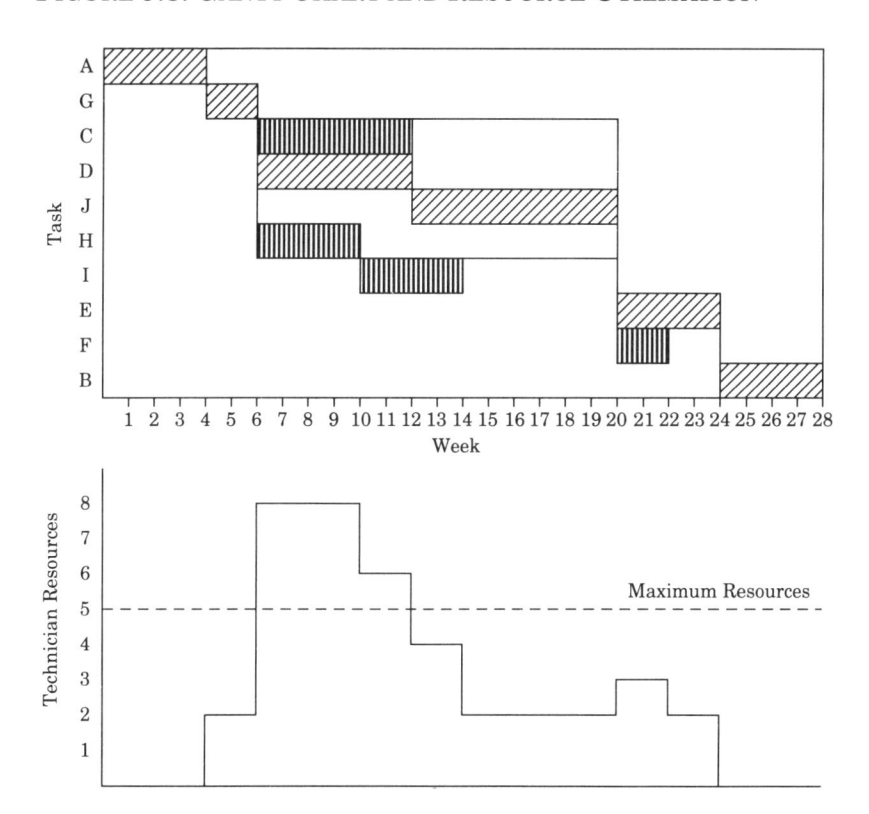

The utilisation of resources for each week of the project is shown by graphing resource requirements against the project time period, as Figure 9.8 above shows.

The graph shows how the utilisation of labour varies for each time period. The graph also clearly shows the tasks giving rise to the labour utilisation. A peak demand for labour occurs during weeks 6 to 12. For efficient utilisation of resources, an even loading or demand is recommended. To achieve an even load, some of the work from weeks 6 to 12 should be transferred to either weeks 1 to 6 or weeks 12 to 28.

Optimising Resource Utilisation

Optimisation of resource utilisation means efficiently distributing demand for resources by levelling the load. Sometimes the level of resources available at any one time is constrained and the project team must schedule tasks to accommodate these constraints. The floating time inherent in non-critical tasks is used to reschedule their start times away from peak demand periods. In the above example, task C is rescheduled to start on week 15 and the graph is redrawn (Figure 9.9). This results in a reduced peak demand for technical labour.

FIGURE 9.9: OPTIMISED RESOURCE UTILISATION

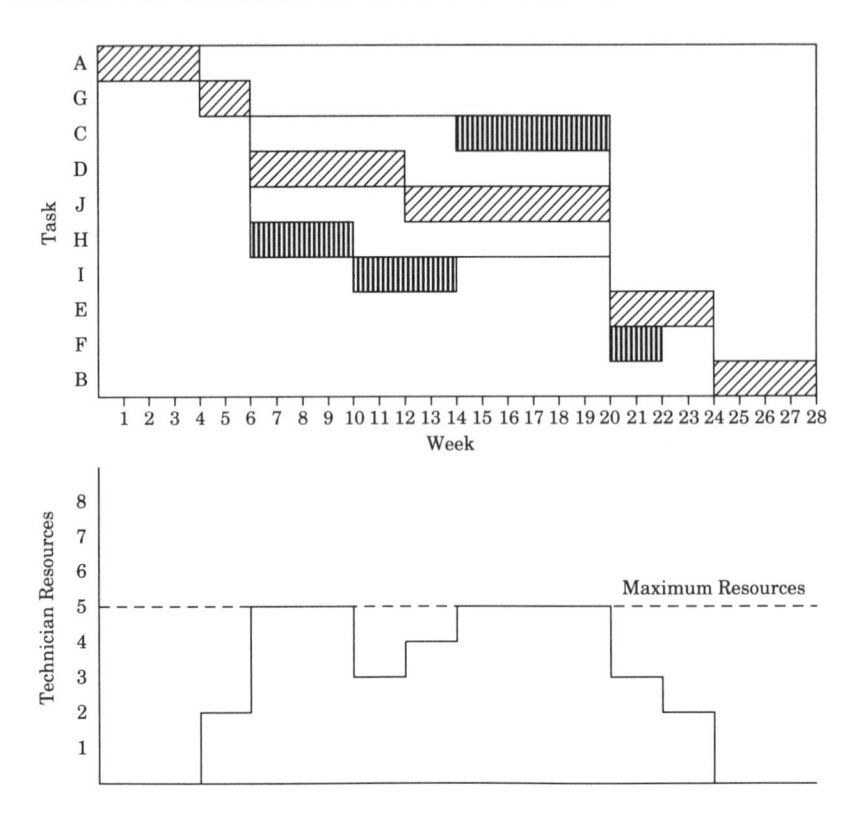

EXPLORING THE TIME/COST RELATIONSHIP

Project costs can be graphed in a similar fashion to that described for other resources such as labour. However, in the case of costs, a cumulative graph probably provides the best information. For each task, the team identifies the financial resources required as follows:

Project Management

Task	Financial Resources (1,000s)
A	5
G	15
C	25
D	20
J	18
H	24
I	27
E	16
F	10
B	8

FIGURE 9.10: CUMULATIVE GRAPH OF PROJECT COSTS

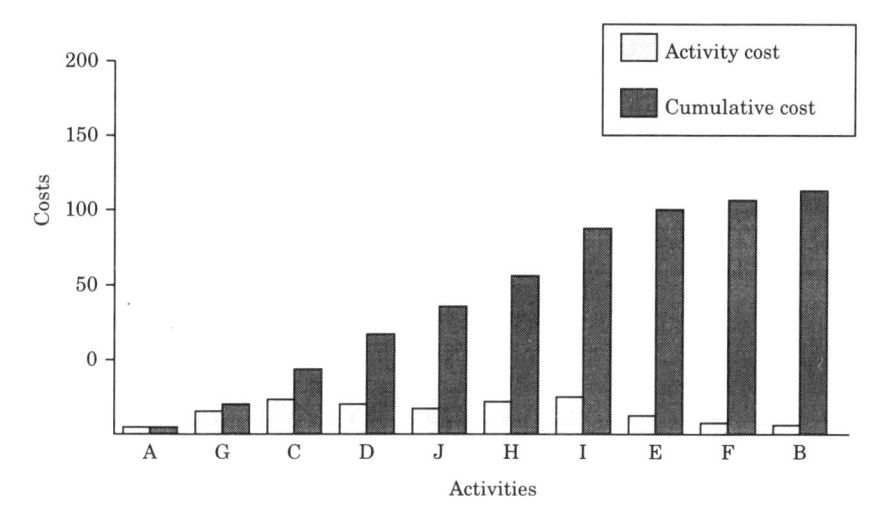

The graph enables the team to readily recognise variation in costs during the project and to predict their effect on the overall project cost. There may be opportunities for compressing the project's overall duration time by spending more money. Very often the estimates for task durations results in an unacceptably long project duration time. The overall project time can only be compressed by reducing the time taken for tasks on the critical path. In some cases, these duration times can be reduced at a cost. The extra cost incurred has to be compared to the cost benefits from time saved. For example, a hotel may want to open its renovated sector for the holiday season and may therefore be willing to pay extra for a shorter project timespan. Another example could be a company launching a new product; it may be critical to shorten the time to market for the product in order to gain first mover advantage.

The team has to identify potential time reductions and their associated costs for tasks on the critical path. Those critical tasks with scope for significant reduced duration at a minimum extra cost are the ideal candidates for further investigation. In some cases, the team may decide to lengthen the schedule in order to reduce costs. Again, the team evaluates the utilisation of money, time and other resources and calculates an optimum based on constraints.

Establishing Milestones

The progress of the project is easier to manage if milestones are identified for the completion of the work at various stages of the project. Assuming the milestones are achieved on time, they serve to confirm that the project is progressing according to schedule. The milestones act as mini-targets for the team, helping to focus effort on short-term targets within the total project time.

Milestones should represent clearly identifiable progress. They should be specific and clearly defined. They mark the path to project completion and, assuming they are achieved on time, the project should be on schedule. In the example below, milestones are represented on the Gantt chart as black squares. The completion of tasks G, (C, J and I), and (E and F) have been identified as milestones.

Project Management   309

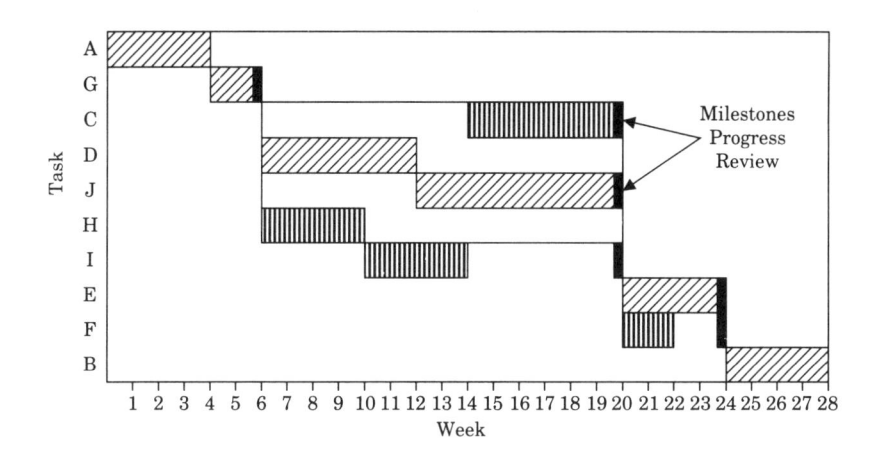

FIGURE 9.11: MILESTONES FOR THE PROJECT

[Gantt chart showing tasks A, G, C, D, J, H, I, E, F, B plotted across weeks 1–28, with Milestones/Progress Review indicated]

Responsibility Matrix

There should be no ambiguity as to who is responsible for the completion of various tasks within the team. A responsibility matrix clearly distributes responsibility throughout the group. The following responsibility matrix links various team members with the project tasks.

FIGURE 9.12: RESPONSIBILITY MATRIX

Activities	Ann	John	Mike	Sofia	Mary	Ted
Task 1	P		S	S	S	
Task 2			P		S	S
Task 3	S		P	S	S	
Task 4	S	S				P
Task 5			S	S	S	P
Task 6	S			P	S	S
Task 7	P	S				S
Task 8		S	S		P	
Task 9	P		S		S	
Task 10		S		S		P

Primary Responsibility = P   Secondary Responsibility = S

The team will need to develop status reports that communicate progress to a wider sphere of people. These reports will show progress against the project's Gantt chart, real costs incurred against forecasted costs, and other useful data. The review reports should emphasise any variation from the original plan in order to ensure maximum effort is applied to meet the overall objective.

Project Management Software

The good news is that there is an abundance of software packages that perform all the mathematical calculations described and generate graphs accordingly. These packages enable the team to perform "what if" analysis with the minimum of effort and exposure to miscalculation. As the project progresses and times vary, the software packages enable the team to easily update the plan.

The software packages on the market are able to generate network diagrams and Gantt charts and portray resource utilisation. The low cost of these packages makes them highly recommendable to any project team.

Implementation Stages

Risk and Contingency Plans

Project failure is something most project teams are reluctant to consider, but early consideration can lead to the development of contingency plans that can save the company and team much grief. There are high levels of risk associated with many new projects. New technologies are expensive and can be difficult to implement successfully, even for the experts. Companies may be staking their reputations and future on the successful conclusion of projects and therefore should seek to reduce risk through contingency planning.

For example, a company may commit to shipping large volumes of units to customers, based on the introduction of new technology. If problems arise with the introduction of the technology and delivery dates cannot be met, the company could lose its customer base. The team can minimise this threat by identifying alternative means of temporarily achieving the company's goals, should the project fail. The team could run a second plan in parallel as a contingency. It may involve developing a subcontractor or a less risky, low-cost manufacturing technology. Ultimately, the team can save much money in developing a contingency plan, depending on the risks associated with project failure.

Resistance to Change

Most projects involve introducing change to the work environment. The change may be perceived as positive or negative by the people affected. Factors encouraging change may include a positive, career-oriented workforce. Other positive factors promoting change may be market expansion or contraction, availability of new technology or new government regulation, for example on health and safety, among many others. Factors against change may be the lack of finance, the conservative culture within the organisation, employees' fear of the unknown and the fear that technology will replace people. Focusing on these factors and addressing these issues can help the team to create an environment conducive to change. These factors play a major role in evaluating the feasibility of the project. The probability of success of any project is greatly enhanced when these obstacles to change are overcome within the project implementation plan.

Kurt Lewin developed force-field analysis as a means for understanding the sources of resistance and the drive for change. Lewin considered that in any situation there are forces driving change and forces restraining change. When the restraining forces exceed the driving forces, the status quo prevails, but when the driving forces exceed the restraining forces, the changes are implemented. The method is subjective rather than numerical by nature, but it helps to clarify obstructions to change.

FIGURE 9.13: FORCE FIELD ANALYSIS

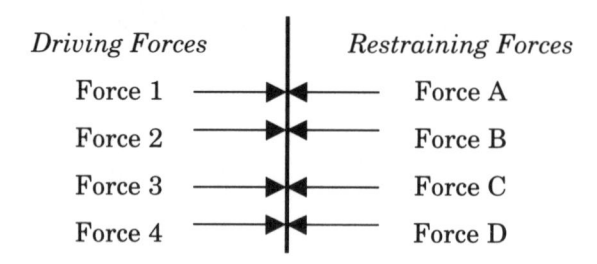

Driving Forces		Restraining Forces
Force 1	→│←	Force A
Force 2	→│←	Force B
Force 3	→│←	Force C
Force 4	→│←	Force D

The idea can be explained by applying it to the example where the general manager wants to introduce ISO 9000 into his facility. A force field analysis lists forces which may prevail:

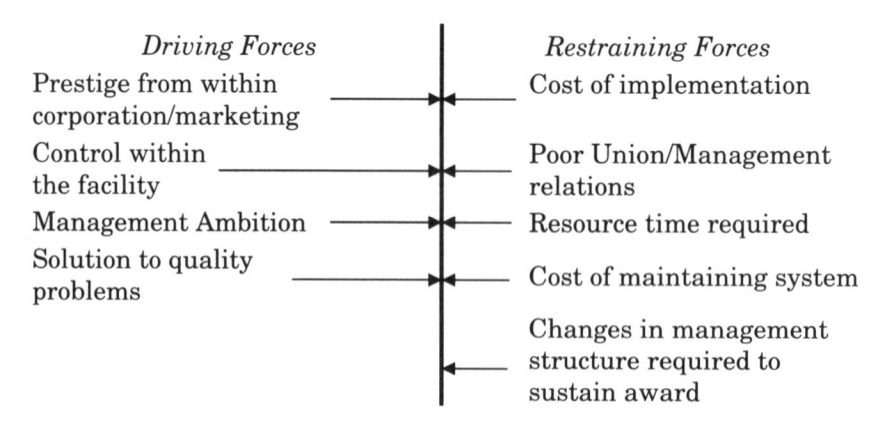

Driving Forces		Restraining Forces
Prestige from within corporation/marketing	→│←	Cost of implementation
Control within the facility	→│←	Poor Union/Management relations
Management Ambition	→│←	Resource time required
Solution to quality problems	→│←	Cost of maintaining system
	│←	Changes in management structure required to sustain award

Listing the forces affecting any situation enables project coordinators to understand clearly any trade-offs that may exist within the project. The sources of resistance are highlighted and, depending on their strength, the driving forces may need to be enhanced or the restraining forces may need to be weakened. Analysing situational forces in this way helps to avoid frustrating, futile attempts at change at an early stage. The approach adopted in altering the situational forces will depend on the time allowed for the transformation to occur and the level of change involved.

SELECTING THE PROJECT TEAM

Project team selection may be the most important task performed in the life of the project, since good choice of team members provides the skills, authority, expertise and, most importantly, commitment needed to succeed. It is very important for the managers of the team members to appreciate the time commitment likely to be required and to enable them to make this commitment. It is also important that team members are not just nominated by their department head as a representative, but that they themselves want to participate in the project. A good team spirit encourages open discussion, innovative ideas and effective decision making.

The size of the team is important. Small teams of three to five people tend to be more suitable where a lot of uncertainty exists on how to proceed. In this environment, there will be a lot of discussion and interaction. A small team may also be suitable where strategic planning of the project is needed. For example, one company that I know, which was implementing a JIT system for the first time on one of its production lines, had a small project team consisting of the area manager, the industrial engineer, the mechanical engineer, the electronic engineer and the production supervisor. The team had to generate ideas on how to resolve particular problems and had to manage an implementation plan that was very uncertain and risky.

A larger team means that more skills and expertise are available to the project. A broader range of people also have an opportunity to participate and take ownership for the success of the project. However, excess debate at these meetings may lead to less effective decision-making.

The effective use of a large team can be seen from the following example. A factory was expanding its production facilities and formed a team to manage the project. The team members were the senior cost accountant, senior human resource administrator, a number of senior engineers, a project planner, industrial engineer, senior systems analyst, production manager and senior quality engineers. Since all the stakeholders in decision-making were part of the implementation team, it was possible to make these decisions quickly. The accountant readily provided information to direct engineering on the cost implications of decisions taken. The quality engineers were able to influence the technology and equipment choices and to ensure a comprehensive approach to product and new equipment qualification. The industrial engineer and the production manager ensured that the layout for the new lines would be appropriate and that the new production areas would be manageable. The senior systems analyst was able to modify production control systems in preparation for the new product lines. The human resource administrator, who would have to hire extra people, learned about the skill levels and the training that would be required.

Team members may also be chosen as a tactic for reducing resistance to change. Sometimes, the concerns and worries of those who resist change can be overcome when they participate in designing the changes. By having some of them on the team, their concerns can be addressed in the development of the project, many pitfalls can be avoided and a broader range of people take responsibility for the success of the project.

CHANGE MANAGEMENT

The path to success is seldom straight and there are many pitfalls that can result in failure. A well thought-out plan is only the first step to project completion. Project implementation frequently affects a wider circle of people beyond the project team, especially manufacturing projects. The team needs to develop a change strategy that takes the concerns of these people into account. Their support of the project may be a vital ingredient for its success. Many projects, such as the introduction of Material Requirement Planning or Just-in-Time, have failed because the project team did not consider how the implementation of the project would affect this broad range of people. Implementing projects usually results in a change in the way people currently perform tasks. In situations where the project affects people outside the project team, a change strategy is required for introducing the new methods.

Kurt Lewin identified three stages to successful change management:

- Stage 1: Unfreezing
- Stage 2: Change
- Stage 3: Refreezing.

Lewin considers that, in order to gain acceptance and support for change, these three steps must be followed. His approach encourages a methodical introduction of the changes required.

Stage 1: Unfreezing

Unfreezing involves preparing people for the expected changes. People should be told about the need for change and what its anticipated results are likely to be. Where possible, the machine operators or technicians who will have to make the changes work should be given an opportunity to participate in the design of the project. Changes such as the introduction of new equipment will depend on technicians and machine operators overcoming any initial teething problems. Many useful ideas can be generated at an early stage of the project through discussion and many obstacles avoided. Participation in the development of the project encourages ownership of the project on the factory floor. It also helps reduce any fears and creates a positive attitude towards the proposed changes. Unfreezing creates the correct environment for introducing change; it prepares people for what is required.

Stage 2: Change

During the change phase, the project plan is fully implemented. Progress is monitored throughout and status reports are used to communicate current performance to plan. The team may suffer from several problems — for example: promised resources which are not provided; senior management support on key decisions which is not forthcoming; a realisation that the timetable for completion of tasks is too tight; or the identification of many problems which were not accounted for in the original plan. The team will have to evaluate each of these events and assess its impact on the viability of the project. Deficiencies in the original plan should be detected at the earliest possible stage so that remedial action can be taken. As mentioned earlier, project planning should include developing a contingency plan where possible. In the event of excessive obstacles in the main plan, the contingency will enable some of the critical objectives to be met.

The change phase describes the actual implementation, when people learn to work with the changes. They adjust their working habits to accommodate the new systems. It may be a turbulent time for some employees who are uncomfortable with change, and management should be aware of the ramifications of the change imposed by the project.

Stage 3: Refreezing

The refreezing phase involves monitoring and consolidating the changes that have been made. The refreezing aims to incorporate the changes into the company's procedures and systems so that they become part of standard work practices. Ideally, the project should encourage a continuous improvement ethos after the initial aims have been accomplished. In this way, the company keeps abreast of the competition and people become familiar with and less fearful of change.

Refreezing ensures that employees do not revert back to the old methods. The team reaffirms the benefits of the new methods and works to resolve any remaining problems. Refreezing engrains the new system as part of the normal work environment.

MANAGING THE IMPLEMENTATION

Managing the implementation involves measuring progress against the plan, identifying variations from the plan and taking corresponding corrective action. The team will have to decide how often it should meet to review progress and how best data relating to progress should be gathered and reported.

The plan is a live document; tasks may get delayed or new information may come to light that may affect the project plan. These variables will all have to be planned for and if, for example, the completion date is affected, the team will need to apply problem-solving techniques to generate ways of overcoming the problems.

Status reports should be generated at regular intervals to communicate progress. These reports should demonstrate how actual progress compares to the original plan. Any exceptions that will affect the overall time or resource utilisation should be highlighted as soon as possible. Regular reviews and status reports ensure that the momentum is maintained and any deviation from the plan is immediately addressed.

CONCLUSION

In today's world of constant change, companies are struggling to remain up-to-date and competitive. Mastering the skills involved in change management greatly facilitates the implementation of those techniques needed to maintain competitive advantage. These skills are networking techniques such as critical path analysis, resource optimisation using Gantt charts, and situational analysis methods such as Lewin's forcefield analysis and the three stages of change.

More than ever before, the effectiveness of manufacturing management is dictated by the ability to handle change. A variety of new technologies and techniques have emerged in the last ten to twenty years, and the success of Western manufacturing depends on how effectively, management can evolve their production facilities to take advantage of these new tools.

Taking advantage of these tools inevitability means introducing change and the best way to introduce change is in a controlled and structured environment through project management. Project management defines the goals, identifies the tasks and ensures that a project team implements the necessary change in a professional and planned manner. Project management is a universal tool which, when implemented, ensures success and sustainability in an increasingly competitive environment.

APPENDIX A

Formulae for calculating Upper and Lower Control Limits:

$$UCL = \bar{\bar{X}} + A.\bar{R}$$

$$LCL = \bar{\bar{X}} - A.\bar{R}$$

where:

$\bar{X}$ = the sample average

$\bar{\bar{X}}$ = the average of the sample averages

R = the range within the sample

$\bar{R}$ = the average of the sample ranges

A = variable

$Cp = (USL - LSL) / (6\sigma)$, where $\sigma = \bar{R}/d2$

Table of Factors for A and d2:

Sample size	A	d2
2	1.880	1.128
3	1.023	1.693
4	0.729	2.059
5	0.577	2.326
6	0.483	2.534
7	0.419	2.704
8	0.373	2.847
9	0.337	2.970
10	0.308	3.078

APPENDIX B

The aim of statistical inference is to draw conclusions about the statistical characteristics of a population from data provided by samples from the population.

Statistical inference consists of two steps. Firstly, a hypothesis is stated regarding characteristics of a population. Evidence regarding the validity of the statement is provided by the statistical characteristics of sample data taken from the population. The hypothesis may be rejected or not rejected with varying degrees of certainty or confidence.

For example, an engineer wants to reduce solderballs within an electronic reflow process. The average failure rate due to solderballs within the process is 0.041 with a standard deviation of 0.001. The engineer changes the process settings and wants to know if the process has now improved. After the change the reject rates for samples are 0.042, 0.038, 0.041, 0.039. Can the engineer say with 95 per cent confidence that the process has improved?

A hypothesis is stated that there is no statistically significant difference between the original performance of the reflow process and its current performance. This hypothesis is generally called the null hypothesis and its validity can be tested to determine if it should be rejected.

A hypothesis should be tested by analysing the sampling distributions of means. The graph of a large number of sample means is a normal distribution; therefore 95 per cent of the sample means will be within +/−2 standard deviations of the population mean. The hypothesis can be rejected with 95 per cent confidence if the mean being tested is outside +/−2 standard deviations from the population mean before the process was changed. If the new mean is outside this 2 standard deviations limit, then it can be said that performance of the current process setting differs from the earlier process setting. The null hypothesis is rejected because there is statistical evidence at the 95 per cent confidence level that the process performance has changed.

On the other hand, the analysis may reveal that the new sample mean is within 2 standard deviations of the mean performance of the process before the changes. In this situation, there is no statistical evidence to reject the hypothesis.

Thus a statistical inference results in a rejection or a non-rejection of the hypothesis under consideration. A non-rejection verdict is not the same as proof of the veracity of the hypothesis; it merely indicates that there is no statistical data that rejects the statement at the prescribed confidence level.

In the above example, it is assumed that a very large sample size was available to describe the population before and after the changes. However, very often only a small sample size is available; in such cases, assumptions enabling the normal distribution to be used for analysis are no longer valid. Instead a t-distribution, a modified version of the normal distribution, is used. The sample size is accounted for by calculating the degrees of freedom. In this context, the degrees of freedom can be considered as equal to the sample size minus one. For example, a sample size of 8 implies 7 degrees of freedom. The t-distribution values are determined from tables of t-values, which have the degrees of freedom on the left hand side of each row and the various levels of confidence at the head of each column. The t-value is called the standard error and its value at 95 per cent confidence level is denoted as $t_{0.95}$.

With large samples, the hypothesis is tested to investigate if the new mean is equal to the old population mean +/–2 standard deviations.

With small samples, the hypothesis is tested to investigate if the new mean is equal to the old population mean +/–$t_{0.95}$ × 2 standard deviations.

Using this alteration, a hypothesis can be tested as described earlier.

Pairwise Comparisons

Pairwise comparison is an extension of the concept of hypothesis testing. When pairs of experiments are performed, such

Appendix B

that one half are carried out under the influence of the factor set at level 1 and the other half are being carried out with the factor set at level 2. For example, the engineer may run five different product types through the reflow process with the old settings, change the oven settings and run another batch of five product types. The engineer will want to know if variability in output is due to inherent variation in performance between the product types or if it is due to the changed reflow settings. If the changes to the reflow settings have no effect on the process performance, then ideally, the difference in output performance between each product type should be zero.

A hypothesis is proposed that there is no difference between the performance of the old process and the new process. This is tested by investigating if the actual mean differs significantly from zero. If the difference is significant at 95 per cent confidence level, then the actual mean is:

Actual mean > +/–t0.95 standard deviation

and the hypothesis is rejected. If the investigation shows that:

Actual mean < +/–t0.95 standard deviation

then there is no statistically significant evidence to reject the hypothesis at 95 per cent confidence.

BIBLIOGRAPHY

Automotive Industry Action Group (AIAG) (1990), *Measurement Systems Analysis: Reference Manual*, Southfield, MI: AIAG.

Automotive Industry Action Group (AIAG) (1991), *Fundamental Statistical Process Control: Reference Manual*, Southfield, MI: AIAG.

Automotive Industry Action Group (AIAG) (1993), *Potential Failure Modes and Effects Analysis*, Southfield, MI: AIAG.

Automotive Industry Action Group (AIAG) (1993), *Production Part Approval Process*, Southfield, MI: AIAG.

Automotive Industry Action Group (AIAG) (1994), *Advanced Product Quality Planning and Control Plan*, Southfield, MI: AIAG.

Automotive Industry Action Group (AIAG) (1994), *Quality System Assessment*, Southfield, MI: AIAG.

Automotive Industry Action Group (AIAG) (1994), *Quality System Assessment Training*, Southfield, MI: AIAG.

Automotive Industry Action Group (AIAG) (1994), *Quality System Requirements: QS 9000*, Southfield, MI: AIAG.

Baker, P. (1996), "ISO 9000: Are Things Getting Any Better?", *Manufacturing Management*, September.

Beatty, C. (1992), "Implementing Advanced Manufacturing Technologies: Rules of the Road", *Sloan Management Review*, Summer, 49–60.

Beitz, W. and Küttner, K.-H. (1994), *Dubbel Handbook of Mechanical Engineering*, London: Springer-Verlag London Ltd.

Bertsche, D., Crawford, C. and Macadam, S.E. (1996), "Is Simulation Better than Experience?", *The McKinsey Quarterly*, No. 1, pp. 50–57.

Bowen, D.E. and Lawler, E.E. (1992), "The Empowerment of Service Workers: What, Why, How and When", *Sloan Management Review*, Vol. 33.

Chrysler Corporation (1990), *Packaging and Shipping Instruction Manual*, Auburn Hills, MI: Chrysler Supplier Quality Office.

Chrysler Corporation (1990), *Shipping/Parts Identification Label Standards Manual*, Auburn Hills, MI: Chrysler Supplier Quality Office.

Costanza, J.R. (1994), *The Quantum Leap in Speed to Market: Demand Flow™ Technology and Business Strategy*, Denver, Colorado: Jc-I-T Institute of Technology.

Crosby, P.B. (1979), *Quality is Free: The Art of Making Quality Certain*, New York: McGraw-Hill.

Cummings, T. and Blumberg, M. (1987), "Advanced Manufacturing Technology and Work Design", in Wall, T.D., Clegg, C.W. and Kemp, N.J. (eds.), *The Human Side of Advanced Manufacturing Technology*, New York, NY: John Wiley and Sons Ltd.

Drucker, P.F. (1959), *The Practice of Management*, London: Heinemann.

Fitt, P.M. (1951), "Engineering Psychology and Equipment Design", in Stevens, S.S. (ed.), *Handbook of Experimental Psychology*, New York, NY: John Wiley and Sons Ltd.

Ford Motor Company (1987), *Statistical Process Control: Instruction Guide*, Dearborn, MI: Ford Motor Company.

Ford Motor Company (1990), *Planning for Quality*, Dearborn, MI: Ford Motor Company Total Quality Excellence and Systems Management, Corporate Quality Office.

Ford Motor Company (1992), *Ford Instructional Systems Design Process*, Dearborn, MI: Ford Motor Company, Instructional Methods Section.

Ford Motor Company (1993), *Failure Modes and Effects Analysis Handbook*, Dearborn, MI: Ford Motor Company, Engineering Materials and Standards, Technical Affairs.

Ford Motor Company (1993), *QOS Assessment and Rating Procedure*, Farmington Hills, MI: Ford Motor Company Quality Publications Department.

Ford Motor Company (1994), *Quality Operating Systems*, Dearborn, MI: Ford Motor Company Supplier Quality Engineering.

Forester, T. (ed.) (1989), *Computers in the Human Context: Information Technology, Productivity and People*, Oxford: Basil Blackwell.

Gitlow, H., Gitlow, S., Oppenheim, A. and Oppenheim, R. (1989), *Tools and Methods for the Improvement of Quality*, Irwin Series, ASQC Quality Press.

Goldratt, E.M. and Cox, J. (1986), *The Goal: A Process of Ongoing Improvement*, Croton-on-Hudson, NY: North River Press.

Groover, M.P. (1987), *Automation, Production Systems and Computer Integrated Manufacturing*, Englewood Cliffs, NJ: Prentice Hall.

Hammer, M. and Champy, J. (1994), *Reengineering the Corporation: A Manifesto for Business Revolution*, London: Nicholas Brealey Publishing.

Herzberg, F. (1966), *Work and the Nature of Man*, World Publishing Co.

Herzberg, F. (1968), "One More Time: How do we Motivate Employees?", *Harvard Business Review*, Vol. 46, pp. 53–62.

Hurley, J.J.P. (1992), "Towards an Organisational Psychology Model for the Acceptance and Utilisation of New Technology in Organisations", *The Irish Journal of Psychology*, Vol. 13, No. 1, pp. 17–31.

International Standards Office (1987), *ISO 10011–1 1990-12-15 (Guidelines for Auditing Quality Systems, Part 1)*, Geneva: ISO.

International Standards Office (1987), *ISO 10011–2 1991-05-01 (Guidelines for Auditing Quality Systems, Part 2: Qualification Criteria for Quality Systems Auditors)*, Geneva: ISO.

International Standards Office (1994), *ISO 9001 – 1994 (Specification for Design/Development, Production, Installation and Servicing, Part 1)*, Geneva: ISO.

Jordan, N. (1963), "Allocation of Functions between Man and Machine in Automated Systems", *Journal of Applied Psychology*, New York, NY: John Wiley and Sons Ltd.

Kenigsberg, I.J. and Abella, R.M. (1991), "TQM: An Expanded View of Quality", paper presented at SME Conference, July.

Labich, K. (1994), "Why Companies Fail", *Fortune*, 14 November.

Law, A.M. and Kelton, W.D. (1982), *Simulation Modelling and Analysis*, New York, NY: McGraw-Hill.

Likert, D. (1967), *The Human Organisation: IT Management and Value*, New York, NY: McGraw-Hill.

Lockyer, K. and Gordon, J. (1996), *Project Management and Project Network Techniques*, London: Pitman.

Logothetis, N. (1991), *Managing for Total Quality: From Deming to Taguchi and SPC*, Englewood Cliffs, NJ: Prentice Hall.

McGregor, D. (1967), *The Professional Manager*, New York, NY: McGraw-Hill.

Montgomery, D.C. (1991), *Introduction to Statistical Quality Control*, New York, NY: John Wiley and Sons Ltd.

Mortimer, J. (ed.) (1985), *Integrated Manufacture*, Bedford: IFS Publications/Ingersoll.

Murata, K. and Harrison, A. (1991), *How to Make Japanese Management Methods Work in the West*, Aldershot: Gower.

Nakajima, S. (1988), *Introduction to TPM — Total Productive Maintenance*, Cambridge, MA: Productivity Press.

O'Sullivan, John (1992), "Solutions through People: A Company-wide Approach to WCM at ALPS Electric (Ireland) Ltd.", Euroform E012.

Peters, T. (1994), *The Tom Peters Seminar: Crazy Times call for Crazy Organisations*, London: Macmillan.

Pilz GmbH & Co. (1995), *Safety for Man and Machine: A Guide to Machinery Safety Standards*, Pilz GmbH & Co.

Pirsig, R.M. (1976), *Zen and the Art of Motorcycle Maintenance*, London: Corgi.

Rommel, G., Kluge, J., Kempis, R.-D., Diederichs, R. and Brück, F. (1995), *Simplicity Wins: How Germany's Mid-sized Industrial Companies Succeed*, Harvard Business School Press.

Schonberger, R.J. (1982), *Japanese Manufacturing Techniques: Nine Hidden Lessons in Simplicity*, New York: Free Press.

Shingo, S. (1988), *Non-stock Production: the Shingo System for Continuous Improvement*, Productivity Press.

Taylor, F.W. (1947), *Scientific Management*, New York, NY: Harper & Row.

Thurow, Lester C. (1992), "Who Owns the Twenty-first Century?", *Sloan Management Review*, Spring, pp. 5–16.

Tuckman, B.W. (1965), "Development Sequences in Small Groups", *Psychological Bulletin*, Vol. 63.

INDEX

absenteeism, 16–17, 18
Alaire, Paul, 255
assembly line, 50–51
attribute charts, *see* statistical process control
automated guided vehicles (AGVs), 221, 265–267
automation, 3, 4, 8, 140, 155–158, 161, 257–291
   assembly, 274–277
   benefits, 258–259
   compared with human, 155–156
   "factory of the future", 279–281
   implementation, 286–290
   strategy, 283–286
   suitability, 156–157, 281–283
   see also automated guided vehicles; computer technologies; conveyor systems; design for assembly; Flexible Manufacturing System; machine vision systems; robotics; technology
autonomous maintenance, 131, 136–137, 139–145, 160, 161
   cleaning and inspection, 142
   contamination, 142–143
   general inspection, 143
   lubrication, 143
autonomy, 25, 28, 36–37

bath-tub curve, 151–152
Beatty, Carol, 286, 287, 289, 291
benchmarking, 160
bottlenecks, 29, 50, 159, 172, 208, 295
Bowen, D.E., 12
brainstorming, 8, 54–55, 56, 80, 155, 159, 296
British Standards Institution, 90, 95

capability indices, 71–74, 126, 319
Capek, Karel, 268
cause-and-effect diagrams, 7, 55–56, 60, 80, 155
Champy, J., 241
checklists, 60, 61
checksheets, 52–53
competition, 1, 3, 128, 161, 163–164, 255, 257, 283, 285
computer simulation, 6–7, 8, 199, 215–238
   algorithms, 223, 225
   alternative systems, 232
   analysis work, 225, 231–234
   animation, 215, 235
   applications, 220–221, 224
   assumptions, 228
   credibility, 231
   data collection, 228–230

computer simulation (cont'd)
  definition, 216–219
  discrete event modelling, 222–223
  implementation, 235
  model creation, 230
  objectives, 227
  pitfalls, 236–237
  preliminary work, 224, 226–231
  software packages, 215, 216–217
  statistics, 235–236
  suitability, 219–222
  teamwork, 226–231
  validation, 231
  verification, 231
computer technologies, 246–249, 254–255
  computer-aided design (CAD), 235, 246, 247, 248, 258, 271, 278
  computer-aided engineering, (CAE), 246, 247
  computer-aided manufacturing, (CAM), 246, 247, 248, 258, 271, 278
  computer-aided process planning, (CAPP), 246, 247, 248, 278
  computer integrated manufacture, (CIM), 246, 247, 248, 272, 277–279, 285, 287
  computer numerically controlled (CNC) machines, 272
  shop-floor data collection, 246, 247, 248, 278

conflict management, *see* team dynamics
consultants, 14, 17
conveyor systems, 262–265, 273
  belt, 263, 264
  chain, 263, 264
  overhead, 263, 264
  roller, 263–264
  shuttle, 265
cost-benefit analysis, 17
Cp index, *see* capability indices
Cpk index, *see* capability indices
customers, 41, 92, 187, 194, 209–210, 219, 221, 236
customisation, 277

decision-making, 11, 17–20, 172
  autocratic, 17
  see also empowerment
dedicated production lines, 168–169
delivery, 2
Deming, W. Edwards, 40, 46, 60, 86–87
design for assembly, 274–277
design, product, 70
developing economies, 257–258
distributions, 72, 229
  exponential, 229
  normal, 63, 64–66, 68, 229
  uniform, 229

employees, 3, 9
  morale, 31
  survey, 15–16
  see also empowerment; operators

Index

empowerment, 7, 9–37
  autonomy (level four), 20, 32, 33, 36
  benefits, 22–24
  decisions (level three), 18–20, 32, 36
  implementation, 13–36
  lack of (level one), 17–18
  pilot implementation, 31, 32–33, 35
  pitfalls, 35–36
  suggestions (level two), 18, 30, 31–32, 36
  suitability, 11–13
engineers, 4, 19, 156, 160, 174, 181, 221, 278, 281, 286, 289, 293
  see also equipment
environment, work, 21
equipment, 5, 147, 149, 181, 190–191
  changeovers, 173–174, 186, 187, 205
  downtime, 2, 134, 147–150, 159, 188, 232
  failure, 57–60, 153, 154
  layout, 168–170, 180–182
  life-cycle, 151–152
  maintenance, 8, 16, 25, 129–161
    analysis, 138, 146–158
    definition, 133–134
    implementation, 138–160
    preliminary work, 139–145
    suitability, 137–138
    see also Total Productive Maintenance
  part replacement, 152, 154
  set-up, 170, 171, 186–189, 192–193, 202
  tools, 272
  uptime, 8, 134

facilitators, 34, 45
factory, 6
  layout, 168–170, 174, 178, 180–182, 190–192, 206, 313–314
  transportation in, 170, 180, 181, 202
  see also automation; equipment
failure modes and effects analysis (FMEA), 7, 56–60, 111
feedback, 25, 28–30
Fitt, P.M., 158, 161
flexibility, 16, 167, 258, 273, 282, 283, 285, 291
Flexible Manufacturing System (FMS), 259, 271–273, 278, 281, 287
  configurations, 273
flow charts, 48–51, 230, 244
force-field analysis, 311–313, 317
Ford, Henry, 279, 281
Forester, Tom, 282

Gantt charts, 7, 293, 294, 303–304, 305, 317
gauge capability analysis, 74–75
globalisation, 257
goals, company, 16
Groover, M.P., 284

Hammer, M., 241
Harrington, Joseph, 277–278
Herzberg, F., 22–24, 30

histogram, 60, 62–64, 304, 305

improvement, 34, 49, 143, 155–158, 240, 242, 255
  continuous, 80, 86, 161, 164, 213, 281
  cycle (plan–do–check–act), 46–48
industrial relations, 46
inspection, 44
International Standards Organisation (ISO), 90
inventory, 5, 165–167, 168, 170, 172, 181, 186, 194, 197–205, 207, 212, 232, 242, 259, 265
  buffers, 166, 199–200, 203
  see also Just-in-Time manufacturing; kanban
Ishikawa, Kaoru, 55
ISO 9000, 8, 43, 89–128, 295
  and QS 9000, 126–127
  audits, 106–108
  benefits, 90–91
  documentation, 93, 102–106
    quality manual, 102–104, 127
    structure, 104–106
  implementation, 108–110
  overview, 94–102
  quality requirements, 95–102
  registration to, 94–95
  suitability, 93

Japan, 1, 5, 46, 86, 139, 163–164, 171, 193, 200, 285, 288
job satisfaction, 22–24
Jordan, N., 156, 158, 161

Just-in-Time manufacturing (JIT), 5–6, 8, 138, 159, 163–213, 237, 283, 295, 313, 314
  benefits, 164
  components of, 167, 212–213
  definition, 164–165
  JIT deliveries, 8, 164, 167, 208–212, 213
  suitability, 167
  see also kanban; structured flow manufacturing

kanban, 8, 164, 166, 167, 193–208, 212, 213
  analysis work, 196–197, 198–205
  cards, 193–195, 200–201, 211–212
  communications, 200–202
  cost–benefit analysis, 197–198
  implementation, 205–208
  preliminary work, 197–198
  see also inventory; Just-in-Time manufacturing (JIT); lot size; one-piece flow; suppliers
Korea, 39

Lawler, E.E.
Lewin, Kurt, 8, 293, 294, 314
lot size, 202–205, 206–207, 208, 280, 311

machines, see equipment
machine vision systems, 260–262
maintenance,
  predictive, 132, 137, 150–151

Index

preventative, 132, 137, 144, 148–150, 154–155
  see also equipment; Total Productive Maintenance
management, 3, 4, 12–13, 16, 18–19, 20, 31, 35, 44, 45, 168, 240–241, 257, 293, 317
manufacturing, 1, 2, 7, 18, 29, 35, 41–42, 161, 163, 168, 198, 220, 241, 278, 281, 293
  components of, 3
  cycle time, 165–167, 233
  lead-time, 1, 164, 167, 172, 173, 202, 206, 258, 280
  mass, 1
  see also, structured flow manufacturing
marketing, 280
Material Requirement Planning, 163, 203, 278, 314
materials, raw, 5, 70, 165, 167, 208, 285
  see also inventory
models, *see* computer simulation
Mortimer, J., 248, 283
motivation, 22–24, 25, 30

network diagrams, 298–302
  critical path analysis, 8, 294, 299–300, 304, 317
  float times, 300–302, 304, 306
  PERT, 7, 299
  task duration, 298–300

one-piece flow, 202–205
operating performance efficiency, 134–135, 158, 159

operators, 3, 4, 9, 10, 16, 19, 20, 22, 28, 31, 33, 35, 54, 131–132, 137, 140, 144, 147, 155–56, 158, 159, 168, 181, 183–186, 205, 273, 286, 288
  see also employees; empowerment
overall equipment effectiveness (OEE), 134, 139, 155, 158–160

pairwise analysis, 233–234, 322–323
Pareto diagram, 53
Pirsig, R., 146
plan–do–check–act (PDCA) cycle, *see* improvement
problem-solving, 45, 53, 79, 81, 155, 159, 317
  see also troubleshooting
procedures, 11, 48
process mapping, 244–245, 250, 251, 253
process-oriented production, 169
production
  mini-lines, 168–169, 170, 171–172, 182–183, 190–191, 193, 199
  schedules, 186–188, 250–252
  volume, 281–282
productivity, 8, 155–156, 258
project champion, 36
project management, 7, 293–317
  change management, 293, 311–313, 314–316
    refreezing, 315, 316
    unfreezing, 315
  defining the project, 295–296
  definition, 294

project management (cont'd)
  duration, 301
  identifying tasks, 296–297
  implementation, 310–313, 316–317
  milestones, 308–309
  objectives, 296
  risk, 310–311
  software, 310
  suitability, 294–295
  team, 313–314
  techniques, 294, 296–310
    see also Gantt charts; histograms; network diagrams; resources; responsibility matrix
  time/cost relationship, 306–309

QS 9000, 8, 91, 110–127
  Advanced Product Quality Planning and Control Plan (APQP), 111,
  and ISO 9000, 126–127
  overview, 111
  Production Part Approval Plan, 111
  requirements, 112–126
    Section 1, 112–122
    Section 2, 122–124
    Section 3 (customer specific requirements), 124–126
  suitability, 110
quality, 1, 2, 8, 20, 39, 40, 60, 86–87, 172, 258, 291
  costs, 43–44
  fitness to cost, 42
  fitness to standard, 42

system, 92–93
  see also ISO 9000; QS 9000; Total Quality Management
questionnaire, 82

range graph, 66, 68
reengineering, process, 7, 8, 237, 239–256, 243–253
  analysis work, 246–253
  configurations, 249–253
  definition, 241
  implementing, 249, 254–255
  performance targets, 245–246
  preliminary work, 243–246
  suitability, 242–243
  see also process mapping
Reliability Centred Maintenance, 132, 137, 151–155
resource
  allocation, 7, 8, 295
  optimisation, 294, 305–306, 317
  utilisation, 304–305
responsibility, 20, 21, 33, 252–253, 297
  matrix, 309–310
robotics, 267–271, 273, 282
run chart, 66–68

safety, 259
skills, 20, 84, 157, 172, 185–186, 281
  variety, 25, 26
Sloan, Alfred, 239
Smith, Adam, 239–240
specification, 42, 62, 68, 72–74
  tolerance, 72, 73

Index

statistical process control (SPC), 7, 30, 68–71, 111
   attribute data, 77–79
   lower control limit (LCL), 68, 69, 70, 319
   upper control limit (UCL), 68, 69, 70, 319
statistical inference, 321–323
statistical techniques, 60–78
structured flow manufacturing, 8, 164, 167–193, 205, 212, 213
   benefits, 173–174
   components of, 170
   costs, 174–175
   group technology, 170, 171–172, 175–178, 189–190, 272
      coding systems, 177–178
      process flow analysis, 176–177
   five S, 170, 171, 178–180, 190
   housekeeping, 170, 171, 178–180, 190
   implementation, 172–191
      analysis work, 172–173, 175–189
      pitfalls, 189–193
      preliminary work, 172, 173–175
      recommendations, 189–193
      workload, balancing, 182–186, 192
   see also equipment; Just-in-Time manufacturing
suggestion scheme, 18, 25
supervisors, 19–20, 28, 33–34, 54, 140, 206
suppliers, 6, 89–90, 91, 194–196, 208, 209–212

task identity, 25, 26–27
task significance, 25, 27–28
tasks,
   improvement, 130, 131, 132
   project, 296–297, 298–302
   simple repetitive, 130, 132, 155, 156
   technical repetitive, 130, 131, 132
Taylor, Frederick, 3, 22
team dynamics, 80–86
   conflict, 82–83
   forming, 81–82
   norming, 84–85
   performing, 85–86
   storming, 82–84
teamwork, 9, 14–15, 19, 34, 36, 47, 71, 80–86, 141, 159, 172, 226–231, 246, 252, 295, 298
   culture, 86
technicians, 130, 131, 132, 147, 148, 155, 158, 160, 181, 243, 253, 286, 288
technology, 5, 7, 129, 155, 240, 257, 280
   see also automation; computer technologies
throughput, 233–234
   see also automation
Total Productive Maintenance, 5, 8, 131–132, 134–137, 139, 142, 160, 161, 283
   see also equipment maintenance
Total Quality Management (TQM), 4, 7, 39–87, 283
   definition, 41
   goals, 45

Total Quality Management (cont'd)
  implementation guidelines, 79–80
  introducing, 44–46
  statistical techniques, 60–79
  tools, 48–60
training, 14, 17, 30–31, 79, 283
troubleshooting, 146–158

United States, 90
U-shaped layout, 181, 184–185, 192

variation, 40, 62, 66, 69, 74, 87, 215–216, 222
  common cause, 70–71
  special cause, 71

waste, elimination of, 164
Western industry, 163–164, 255–256, 317
work breakdown structure, 296–297
workload, 130–132, 156, 182–186, 192